# WILD HEALTH

# WILD HEALTH

Lessons in Natural Wellness
from the Animal Kingdom

CINDY ENGEL

HOUGHTON MIFFLIN COMPANY
BOSTON · NEW YORK

First Houghton Mifflin paperback edition 2003

Visit our Web site: www.houghtonmifflinbooks.com.

Library of Congress Cataloging-in-Publication data is available.

ISBN 0-618-07178-4
ISBN 0-618-34068-8 (pbk.)

Printed in the United States of America

Book design by Robert Overholtzer

QUM 10 9 8 7 6 5 4 3 2 1

For Selva and Wilf — Be happy.

# ACKNOWLEDGMENTS

I thank my editor, Harry Foster, for his enthusiasm and commitment to this book and for his work in shaping an amorphous mass into a readable script. I thank Professors Richard W. Wrangham of Harvard University and Michael A. Huffman of Kyoto University for their encouragement and for sharing their observations of chimpanzee self-medication in Tanzania. Mike has been exceptionally supportive, and for this I am deeply grateful.

Many thanks to the many people who took time out of their busy schedules to provide detailed answers to my questions: Michael Appleby, Steve Appleyard, Mary Baker, Phil Baker, John Berry, John Bradshaw, Jason Byrd, Dale H. Clayton, Orin Courtenay, Donald Cousins, Ellen Dierenfeld, Anna Dixon, Holly Dublin, Robert Dudley, Greg Finstad, Murray E. Fowler, Robert Furness, Bennett G. Galef, Jr., Kenneth Glander, Matthew Gompper, Harry Greene, Helga Gwinner, Bob Harding, John Hare, Warner Jens, Timothy Johns, Marci Johnson, Steve Kestin, Jean-Martine Lapointe, David S. Maehr, William C. Mahaney, Georgia Mason, Sarah Mason, Chris Mead, David Mech, Torsten Mörner, R. Norgren, S. Piers Simpkin, Joyce Poole, Tim Roper, Mike Samuel, Ronald K. Siegel, Reno Sommerhalder, Philip Starks, Mikkel Stelvig, Douglas Tallamy, Bryan Turner, Sylvia Vitazkova, and Trevor Watkins.

I am grateful also to the many herbalists, amateur naturalists, and specialists who contributed their observations — in particular, Neal Bosshardt, Richard Guy, Steve Jackson, Ricardo Leizaola, Jane Pointer, Chris Redenbach, Bill Roundy, Gilles Toanen, and Wendy Volhard. The Association for the Study of Animal Behaviour, the International Bear Association, and the Animal Behavior Society were kind enough to publish my call for information.

I thank Caroline Davidson for being so much more than a literary agent, and Kathleen Anderson for her efforts on my behalf in the United States. John A. Burton, Bernardine Freud, Jill Reece, and Martin Wolfe gave polite feedback on early drafts, and the staff of Halesworth Library, Suffolk, made unceasing efforts to find highly obscure documents.

And although I have not personally met the animal herbalist Juliette de Baïracli Levy, I am indebted to her years of observation, which inspired my interest in animal self-medication. Finally, I acknowledge the contribution of the variety of wild animals that demonstrated their "behavioral strategies" to me at opportune moments.

# CONTENTS

## III. LESSONS WE MIGHT LEARN

Ask now the beasts, and they shall teach thee;
and the fowls of the air, and they shall tell thee.

—Job 12: 7

# INTRODUCTION

The dog taken by fever seeks rest in a quiet corner, but is found
eating herbs when his stomach is upset. Nobody taught him
what herbs to eat, but he will instinctively seek those that make
him vomit or improve his condition in some other way.

— Henry Sigerist, American physician, 1951

A sick animal retires to a secluded place and fasts until its body
is restored to normal. During the fast it partakes only of water
and the medicinal herbs which inherited intelligence teaches it
instinctively to seek. I have watched . . . self-healing so often.

— Juliette de Baïracli Levy, European traditional herbalist, 1984

When the kunkis [tame elephants] are sick, the mahouts take
them to the forest where the elephants pick the herbs or plants
they need. Somehow they're able to prescribe their own
medicine.

— Dinesh Choudhury, Indian elephant hunter, 2000

OBSERVATIONS OF ANIMALS healing themselves with natural remedies have been documented in a long line of chronicles stretching
back through Medieval Europe to the ancient civilizations of Rome,
South America, and China. Numerous plants are named after animals that appear to use them in this way. The Navajo and Blackfeet
Indians of North America watched wild bears dig up and use the
roots of *Ligusticum* plants so frequently, and with such obvious benefit, that they named the plant "bear medicine." The Chippewa Indians in Canada call *Apocynum,* one of *their* strongest herbs, "bear

medicine." European herbalists saw how sick dogs turned to eating a particular species of grass (*Agropyron repens*) so reliably that they called it "dog grass." Other examples include hare's lettuce, pigweed, catnip, and even horny-goat weed. Folk medicine is often reputed to have its roots in observations of animal behavior, and rural physicians have noted that North American pioneers discovered the rudiments of their folk medicine in the healing plants sought out by sick animals.[1]

Until recently, though, scientists have been reluctant to accept stories such as these, dismissing them as a romantic anthropomorphism that inappropriately assigns human attributes to animals. In some cases this is clearly so; but, as we shall see, scientists themselves have recently recorded animals performing actions that unquestionably look like self-medication, and it is no longer acceptable to dismiss all such stories out of hand. The subject deserves far greater scrutiny by the scientific community. For if wild animals *are* self-medicating, the implications are vast, not only for pharmacists searching for new drugs, but also for wildlife custodians hoping to improve the management and protection of wildlife, pet owners wanting to better the health of their companion animals, and farmers seeking to enhance the health of their livestock. We might even find ways of improving human health. As health fads come and go, and the facts appear increasingly contradictory, we are in urgent need of some common sense — some firm ground — on which to found our health care. If we can learn (or relearn) guidelines from the actions of wild animals, we should do so.

So far, much of the popular media response to these scientific findings has sensationalized animal self-medication as an example of supernatural animal wisdom.[2] However, we shall see that there are other, more plausible explanations for how animals manage to keep themselves well.

To investigate animal self-medication scientifically, we must set the behavior in the wider context of general health maintenance. My aim in this book, then, is not only to establish whether animals are indeed self-medicating their ills, but also to explore what actions animals take to keep themselves well: how they deal with injury, infection, parasites, a barrage of biting insects, and even the debility of aging and of psychological disturbances. The process of teasing fact from fiction yields exciting and valuable new information for anyone interested in animal and human health.

The topic is huge. If I were to explore every aspect of how animals stay healthy in the wild, this book would certainly run to many volumes: several on the genetics of health, others on immunity, and still others on the social and psychological dynamics of health. Even limiting myself to the *actions* animals take to stay well, I have had to plow through a vast sea of interweaving material. It is not ideal to separate animal behavior from its associated physiology, as the two are inextricably interconnected (and both are interwoven with the environment of the animal), but the extraction does afford a focus that has hitherto been neglected. What emerges is a mere introduction or overview to a vast field of inquiry.

I should say a little about the animals of this book. Over a million animal species have been named, but estimates of the total number range from 1.5 million to more than 30 million (approximately 90 percent of these are thought to be in the tropics), and by far the majority of the species identified to date are insects. Most of the examples I use are birds and mammals, simply because their behavior has received more attention. Still, I have done my best to balance this bias by including examples of insect, fish, amphibian, and reptilian behavior wherever possible. A *wild* animal is defined as a product of natural selection in which only the individuals most suited to their environment have survived to breed, and the less well suited have been excluded from the gene pool. Thus, a wild animal in captivity is still (genetically) a wild animal. A domestic animal, in contrast, has been artificially selected by humans for certain characteristics such as temperament or appearance. Those individuals with the desired characteristics are used for breeding; those without are not. The domestic animal is therefore very different from its wild counterpart. In addition, there are so-called feral animals that were in the past subject to artificial selection but have since returned to the wild and to natural selection.

The human animal, *Homo sapiens,* is a hominid (human-type primate) that appeared about three hundred thousand years ago, having diverged from its closest primate relatives, the great apes, two million years before that. Throughout much of this book, I write as if we were separate or different in some way from other animals (because in many ways we are). However, it soon becomes apparent that we are not sufficiently separate or different enough to ignore the wealth of health advice available from these studies of wild animals.

My exploration of wild health has been a truly interdisciplinary journey in which overlapping and intertwining threads from specialist fields have been brought together, often for the first time. Into this eclectic mix is interwoven material from folklore, traditional medicine, and experimental and anecdotal observations — lest any shred of useful information be lost in the formation of this new fabric. Sometimes science and folklore reinforce each other. Sometimes they do not. As a biologist, I have many years of field and laboratory experience in the study of animal behavior, yet I am not an expert in any of the specialist fields described herein. This fact may work to my advantage, as the benefits of presenting an overview of wild health far outweigh the costs of losing detail in each area. Specialists may be irritated by my oversimplifications of their particular fields, but I encourage them to persevere for the sake of the emergent vista. Should readers wish to pursue the intricacies of any specific discipline, they will find review articles in the notes, and I welcome any additions or comments via my Web site (www.animalselfmedication.com).

The animal kingdom is replete with examples of animals actively helping themselves to stay well. We would be wise to learn what we can from their actions for, despite the medical advances of the last century, health issues still loom large among our concerns. The wealth of successful strategies for health, created over millennia, offers the potential for providing sustainable health care for animals in our care — and for people around the world.

# PART I
# LIVING WILD

1

# HEALTH IN THE WILD

> The multitude of the sick shall not make us deny the existence of health.
>
> — Ralph Waldo Emerson, 1860

THE HERBALIST Juliette de Baïracli Levy has spent much of her long life observing the way animals keep themselves well in the wild. In one of her many books she writes, "Everywhere in the woods one observes the wild animals rearing their young in health and freedom from sickness."[1] But this view is considered naively romantic by wildlife health experts. Although an animal may *seem* healthy on the surface, it may harbor diseases and parasites that drain its resources and can flare up should resistance falter momentarily. Furthermore, the animals we see are the survivors, disease and death having filtered out the less healthy. The wild animal, from this perspective, fights a perennial battle with sickness and disease.

Which view is correct — the romantic vision of a healthy and harmoniously balanced ecosystem, or the survivalist vision of a ruthless, endless battle with death and disease? Paradoxically, the two views are not as diametrically opposed as they might first appear. When we see a beautiful swan glide across still water, the movement appears effortless; the swan seems calm and untroubled, even serene. An observer below the water, however, would see that the swan is working hard: muscles are contracting and relaxing; legs and webbed feet are pumping, pushing water aside with great effort. So it is with wild health. While an animal may appear to glide effortlessly through life's troubled waters, a continuous struggle for survival goes on, largely unseen. One perspective, then, is that behind a façade of blissful, harmonious balance, each and every organism is working to maintain its health and to survive. Another perspective is that the struggle and selective survival actually *create* the impression of harmony. I find no conflict in being able to see both the struggle and the balance in the same vista, but evidently the answer to the seemingly straightforward question "How healthy are wild animals?" is influenced by the perspective of the observer.

Most of us gain our impressions of health in the wild primarily from the news media — and news about wildlife, like news about anything, is seldom *good* news. Currently, wildlife health makes grim reading. Seal and dolphin populations in the Mediterranean and Baltic seas, and in the coastal waters of the United States, have been seriously affected by major disease outbreaks. It looks as if the butylins used to protect the hulls of ships from barnacles and such are the main culprits. These biocidal chemicals damage mammalian immune systems, lowering resistance to disease and cancers. Meanwhile, harbor porpoises in the English Channel and southern North Sea are sickened by the high concentrations of polychlorinated biphenols and mercury in their waters. And a global epidemic of mysterious tumors affecting endangered sea turtles is linked to the pollution of their watery breeding grounds.

On land, amphibians around the world are facing a health crisis. Over the past two decades there has been a rapid decline in their numbers, including extinction of some species, apparently because of a global epidemic of a particular fungal infection. Furthermore, the number of grossly abnormal amphibians born has increased. Although the exact causes of this crisis are ambiguous, environmental factors that disrupt both disease resistance and the developmental systems of amphibians may play a role. All the main contenders are caused by humans: global warming, agrochemicals, and damage to the ozone layer.

Pollution distorts our impression of wild health, and the occurrence of disease in wild-animal populations has become an important indicator of ecological disruption. For a clearer picture of how animals stay well, we need to assess the health of populations far from the effects of industrial society. But even there our presence can disrupt the survey. Early in the study of wild chimpanzees at Gombe National Park in Tanzania, a polio outbreak decimated the chimpanzees, killing four and leaving six permanently disabled. It is thought that the virus spread from local humans, who suffered a polio outbreak a month before, and was carried inadvertently by vaccinated human scientists. The introduction of new pathogens can be devastating for any population. The Spanish conquistadors of the fifteenth and sixteenth centuries killed most of the native Central Americans, not by superior warfare or cunning intellect, but by bringing with them novel and consequently lethal diseases (measles, for one). Today pathogens are traveling the world with increasing ease as the international trade in food, plants, and animals expands, and humans become increasingly mobile. As a result, wildlife is exposed to many new diseases.

As the human population increases, the need for more and more land for housing, agriculture, and tourism continues to squeeze wildlife into ever-shrinking areas of natural habitat. Asian elephants no longer have enough room to find the food and water they need to stay well. Lions in the Serengeti National Park, along with the last few viable populations of African wild dogs, have been ravaged by canine distemper virus and rabies caught from domestic dogs skirting the edge of the park. William Conway of the Wildlife Conservation Society puts it succinctly: "Our growing herds and flocks of domestic animals have become a plague to wildlife, devastating habitat and spreading disease."[2]

We hear far more about disease passed in the other direction — from wild to domesticated animals. In Europe, wild badgers are blamed by farmers for infecting domesticated cattle with tuberculosis, deer are feared as carriers of foot-and-mouth disease because infected herds can remain symptom free, and wild boar are hounded for spreading classical swine fever (CSF) to commercial pigs because "CSF has become milder in wild boar than pigs."[3] In North America, free-ranging bison are accused of spreading brucellosis to ranched cattle, and wild deer of spreading tuberculosis to cattle. In what I consider to be a totally illogical response, wild animals successfully keeping disease at bay are often killed in order to protect sickly (but profitable) domesticated livestock from infection. In the United

Kingdom, for example, a culling program is currently under way in which twenty thousand badgers will be killed to prevent them from *possibly* spreading tuberculosis to cattle.

This fear of wild animals as harbingers of disease is deeply ingrained in the human psyche. The European hedgehog (small, spiny heroine of a classic Beatrix Potter story) was recently described as "among the most dangerous animals in Europe" by pathologist Ian Keymer of London Zoo, who found that they carry at least sixteen diseases known to affect people. And those "could be the tip of the iceberg," he adds. "If we look closer we may find many more."[4] Howard Hughes would have understood, but if we follow this line of reasoning to the extreme, we should never exchange air or body fluids with other people, and we should certainly eradicate all other species on earth — just to be safe!

Even though wild animals are able to carry diseases that affect livestock and humans, it would seem sensible to explore why they are so successful in fending off the worst effects of these diseases, to look to them for ways of improving our own health and that of our livestock, rather than trying to eradicate them. In addition to looking at genetic resistance to disease, we would do well to learn from the many behavioral self-help strategies that wild animals employ.

One difficulty in assessing wild health is locating genuinely wild places — where animals are not confined by perimeter fences, culled, managed, or exposed to domestic animals or humans. Where is the wild *truly* wild? Unfortunately, such habitats are shrinking daily. A survey by the World Wide Fund for Nature found that more than a third of the planet's animal and plant species exist exclusively on a scant 1.4 percent of its land surface.[5] Moreover, few places on earth remain uncontaminated by persistent pollutants such as PCBs, dioxins, and DDT. With shrinking habitat and increasing pollution, the opportunity to study undisturbed animal populations is decreasing, while the need to do so becomes ever more urgent.

Even when we can find truly wild places, measuring or assessing the health of animals living there is notoriously difficult. Most of the evidence has tended to come from the incidental comments of natural historians and scientists on the health of animals being observed. In the early 1960s George Schaller, of the New York Zoological Society, was the first person to study wild lowland gorillas in West Africa.

He found them healthy, lean, well muscled with shiny coats, although he noted that they did catch cold when the rains came. He was surprised to find roundworms in half the fecal samples he examined because the gorillas were in such good health. During the same time, Jane Goodall found wild chimpanzees to be generally healthy, although they too quite often suffered from colds and coughs during the rainy season. In the 1970s Cynthia Moss started her long-term study of elephants in Amboseli National Park and found them in extremely good health. (Things went wrong in later years, though, when human encroachment and drought struck the herds.) They had few diseases and only a few cases of unexplained sickness. They rarely suffered from contagious epidemic diseases, such as rinderpest, and were able to live to a ripe old age as long as they could avoid drought and human hunters.[6]

Schaller later went to Kanha, in India, where he found disease rare among free-ranging, well-nourished chital (medium-sized deer) and gaur (wild relatives of the cow). He concluded that the health of domestic and wild hoofed animals is mainly a function of the quality of the range, and that animals in poor condition as a result of malnutrition become highly susceptible to parasites and disease. Similar conclusions have been drawn by wildlife veterinarians, who report that free-living marsupials in Australia have few problems with infectious diseases, parasites, and cancers, unless droughts, floods, or range restriction occur.[7]

Unfortunately, anecdotal observations such as these are not adequate to provide an accurate scientific picture of wild health. Sick animals may alter their behavior in ways that make them harder (or easier) to spot than healthy animals. Sick elephants, for example, often separate from the herd to remain near water, shade, and easy food, so an observer might underestimate the prevalence of sickness. Hedgehogs, normally nocturnal, when sick will sit in the sun during the day. A daytime observer might therefore think that hedgehogs were more sickly than they are. More visible populations can be taken as representing a species when this is not necessarily so. Red foxes in the United Kingdom have moved into cities where food is more plentiful and energy rich. These urban foxes live in conditions much more crowded than their rural counterparts, because the food supply is more concentrated. The health of the more visible urban foxes is therefore not an accurate indication of the health of wild foxes in their natural habitat.

Fortunately we are now seeing a minor flurry of health assessments of wild animals living in some of the remotest parts of the world, far from human settlements and pollution. In the 1990s the Wildlife Conservation Society's field veterinary program, headed by William Karesh, ascertained that anacondas in Venezuela, macaws in Peru, rock-hopper penguins and guanacos in Argentina, impala in northern Namibia, forest duiker and pancake tortoises in Tanzania, and African buffalo were all "in good physical condition." They were muscled and lean, had cleanly healed serious wounds, were harboring surprisingly few internal or external parasites, and showed no signs of physical abnormalities such as those currently seen in amphibians. Blood tests revealed that parrots had few infections with common bird diseases and were successfully carrying avian viruses that commonly wipe out captive parrots. Impala had surprisingly few previous infections with local diseases, and duiker carried serious pathogens such as leptospirosis with no visible ill effects.[8] African buffalo were outstanding in their ability to resist disease: "When buffalo encounter viral and bacterial diseases, they generally suffer little." They were in excellent health, yet blood tests revealed that they had been in contact with leptospirosis, parainfluenza, brucellosis, bovine herpes, bluetongue, and foot-and-mouth. As successful combatants of such infections, they were considered "carriers" of disease and much despised by local cattle farmers.[9]

These assessments tend to bear out the earlier observations of Schaller, Goodall, and others that wild animals are often infected with disease-causing organisms (pathogens) *without* showing any symptoms. Repeatedly, animals appear to be in good condition when blood and fecal tests show infection with pathogens or parasites. We have to conclude that it is normal — natural — to be infected with low levels of pathogens and parasites in the wild, but that somehow these are kept below symptomatic levels. Benjamin Hart, a veterinary research scientist at the University of California, Davis, concludes that "wild animals generally are often immune to vector-borne diseases and show few clinical signs of illness from parasite infections."[10]

Whether you consider such animals to be healthy or not depends on whether you think the presence of the pathogen is the same as the presence of the disease. In my view, it is not necessarily the same: to carry pathogens *without* showing symptoms might be considered a sign of extremely good health. Nor does it matter that these animals are only the survivors — that the unhealthy ones simply failed to make it. It is *because* they are survivors that they are of interest to us. How is that they have survived and maintained their health while others have not? They are not merely survivors; they are doing very well. We should be interested in any behavior that has contributed to this condition.

My unsurprising conclusion is that when wild animals are free to range over undisturbed habitat, not exposed to high levels of pollutants and not exposed to extremes of environmental change, they *are* generally in good health. They live within an ecosystem to which their physiology and behavior are, by virtue of their very survival, well adapted. They have been exposed to local pathogens from an early age, so that their immune system is primed (as it would be by vaccinations) for resistance to them. They may get sick; but when they do, the reason is primarily a strong disruption in their environmental conditions (drought, pollution, lack of food, overcrowding, or invasion by a novel pathogen).

Of course, the immune system plays an enormous role in maintaining health, but it is by no means the only line of defense an animal has — and it is certainly not independent of behavior. Scientists at Stanford University captured healthy wild African green monkeys and caged them separately to monitor the effects of stress on their immune systems. The monkeys rapidly succumbed to infectious diseases, and some even died despite being given all the nutrients they were thought to need.[11] This is not an isolated case. It is well documented that healthy wild animals do not take readily to captivity. Immune collapse is common. It is notoriously difficult to maintain the health of wild-born gorillas in captivity even if the animals were healthy when caught. White sharks cannot be held captive at all; they die within weeks.[12] The health of the immune system is demonstrably interlocked with the animal's behavior in its environment.

Good health is therefore a balance between the opposing survival instincts of the individual animal and all the other organisms with which it shares its habitat. The ubiquitous nature of pathogens means that constant attention is needed in order to remain healthy.

Animals cannot simply rely on their immune systems to keep them well under constantly changing conditions. They must take an active role in maintaining their own health. Incredibly, although the physiology and immunology of disease are well researched, the behavioral aspects of health maintenance have not received a great deal of attention. (Benjamin Hart has published the only recent reviews on this subject.)

Paradoxically, animal health research programs rarely study *health*. Like human health research, animal health research concentrates on sickness and disease rather than on how or why certain individuals remain healthy in their natural surroundings. Disease processes are usually studied in domesticated species and under sterile laboratory conditions in which the animal is unable to influence the course of the disease. In other words, the study animals are observed passively enduring disease rather than actively managing their own health. Until recently, scientists have not focused on whether animals have any successful strategies for dealing with disease.

To a layman the terms "health" and "fitness" may be synonymous, but to a biologist they are very different. For biologists, fitness is measured by the number of offspring an animal has (that survive to reproduce themselves) compared to other individuals in its population. An individual animal could, therefore, be considered fitter than another if it successfully reared more offspring — even if it was coughing up blood and dragging itself along on paralyzed limbs! Of course, the fittest animal in a population is often (although certainly not always) strong and healthy as well, but the term "survival of the fittest" is frequently misused to refer to strength or health. Even biologists pay little attention to health, considering fitness a far more relevant attribute. However, the ways animals maximize their health is clearly a pivotal mechanism by which they increase fitness. Survival includes the important aspect of *quality*. Merely surviving is not enough. A wild animal has to survive in *as healthy a condition as possible* in order to compete successfully with others and reproduce. If behaving in certain ways enhances the health of certain individuals, it will give them an adaptive advantage over others that do not behave in those ways. There need be no conscious, deliberate, or intentional basis to these behaviors.

What aspects of animal behavior are we looking for? Basically, any action that quantitatively prevents or treats ill health. There are several possible approaches. An important aspect of all living organisms is that they manage to keep their insides in a fairly steady state with respect to what is on the outside. It was back in 1857 that the French physiologist Claude Bernard discovered that organisms actively strive to maintain their internal state within a narrow range. This phenomenon, called homeostasis, allows the organism a degree of independence from exterior conditions. Many homeostatic mechanisms are physiological: that is, they involve the processes and functioning of the body. As external temperature increases slightly, for example, a mammal might start to sweat; the capillaries near the surface of its skin dilate so that the blood is cooled near the outside of the body and internal temperature is stabilized. But if these physiological changes do not successfully rebalance body temperature, the animal may change its behavior by seeking shade or lying in cool water.

With this vastly oversimplified example, it is easy to see how physiology and behavior interact to keep the internal state "balanced" and thereby maintain health. Many aspects of health maintenance behavior can be described as homeostatic. Other behaviors such as grooming, resting, or even fasting are best described as self-maintenance. Some actions, though, are only taken in response to a health disruption and are called (unsurprisingly) illness response behaviors. Some involve the animal's use of a substance not made by itself, in such a way as to rectify malaise, and are therefore called self-medication. Collectively, these health maintenance behaviors are the focus of this book.

✖ ✖ ✖ ✖

# 2

# NATURE'S PHARMACY

> A plant is like a self-willed man, from whom we can obtain all
> that we desire, if we only treat him his own way.
> — Goethe, 1809

ANIMALS, UNLIKE GREEN PLANTS, cannot manufacture most of the chemicals they need to function. Instead, they must rely on plants to directly or indirectly provide them with the essentials for life. Animal health is therefore intimately dependent on plant chemistry. From the basic materials of sunlight, air, and soil water, green plants manufacture carbohydrates, proteins, fats, hormones, vitamins, enzymes — everything they need to grow, repair damage, and reproduce. In addition to the chemicals they produce for ordinary *primary* metabolism, many plants synthesize so-called *secondary* compounds that seem to serve no obvious metabolic purpose. What is noticeable about these secondary compounds is their biological reactivity — their toxic or medicinal nature — and accordingly they form the bulk of nature's pharmacy. So far, about one hundred thousand different secondary compounds have been discovered.

One explanation for their existence is that they are waste products the plants are unable to excrete. A more widely accepted explanation is that they play a purposeful role as defensive compounds. Plants, like animals, need to protect themselves from infection with bacteria, viruses, and fungi; many secondary compounds work powerfully against these agents. Plants can also produce specific defensive proteins during an infection, comparable to the human immune response. Like our own antibodies, these proteins can provide long-term resistance.

In addition to fighting off disease, plants need to defend themselves against predators (herbivorous insects, mammals, and birds) to which they are quite literally sitting targets. The result has been the evolution of not only physical and structural deterrents such as barbs, stings, spines, and prickles, but also secondary compounds that serve as feeding deterrents. Plants and animals are in an ongoing, never-ending evolutionary arms race. One intriguing fact about plant secondary compounds is that there are generally more of them per plant in the tropics than in temperate regions. Presumably, greater predation in tropical environments requires a larger arsenal of chemical warfare.

Ideally, smell alone should be enough to deter a herbivore from feeding, but if not, then the first taste needs to be sufficiently unpleasant to prevent further damage. The oldest of the chemical feeding deterrents are the condensed tannins, reputed to have defended plants from dinosaurs. They are highly astringent, causing the tongue to pucker and the mucous membranes of the mouth and throat to dry. Once eaten, they continue to cause problems, interfering with digestion by disrupting the environment of important microorganisms and enzymes in the gut. For these reasons tannin-rich plants are usually avoided by grazing animals. Medicinally, though, they are antidiarrheal, antiseptic, antibacterial, anthelmintic (active against intestinal parasites), and antifungal.

Many toxic secondary compounds taste bitter and are eaten only in small quantities by most animals. Saponins, for example, protect a plant against attack from mollusks, insects, fungi, and bacteria. They affect the movement of molecules across cell membranes and can even destroy red blood cells. Alkaloids, produced by about 20 percent of flowering plants (mostly in the tropics), also taste bitter. They are highly reactive compounds that have strong physiological effects on animals even in very low doses. Plants store them in peripheral parts such as bark, leaves, and fruit. Atropine from the belladonna plant, psilocybin from "magic mushrooms," and mescaline from peyote plants are all alkaloids. Hemlock, famously used by European witches, contains eight alkaloids toxic to many animals, including humans. And the bark of the chinchoa tree contains the alkaloid quinine, an effective antimalarial agent. Several plants, including the tobacco plant, contain the well-known alkaloid nicotine. Some alkaloids are remarkably similar structurally to neurotransmitters in the animal central nervous system, including that of humans. Dopamine, serotonin, and acetylcholine are examples. As multipurpose defense compounds they have a broad spectrum of activity. Many are toxic to insects and vertebrates, as well as being able to inhibit the growth of bacteria and other plant seedlings.[1]

Plants are also able to defend themselves with secondary compounds released only when under attack. When acacia shrubs are nibbled by deer, they regrow leaves with higher concentrations of defensive toxins, which cause neurological and reproductive disturbances in any deer unwise enough to return for a second feed. Inducible defenses are called into action against microbial attack, too. When grapevines are attacked by fungi, they release a potent antifungal compound called resveratrol. Found largely in the skin of red grapes, its bioactive effects remain even after the plant has been processed into wine. Its presence is one reason why regular moderate drinking of red wine has beneficial protective effects against heart disease and cancer.[2]

A secondary compound in poison ivy (*Toxicodendron* sp.) causes contact dermatitis, a painful blistering of the skin that can continue for weeks after contact and can even kill particularly sensitive people. Although it is certainly an effective deterrent to most grazing animals, the same plant material has been used by humans to cure bacterial infections such as gonorrhea, dysentery, and gangrene, and it has been shown in laboratory tests to have antimicrobial properties.

Apart from fast-acting feeding deterrents, plants have evolved other more insidious ways of reducing herbivore damage. Some secondary compounds mimic hormones that disrupt herbivore growth and reproduction. The alkaloids caffeine and nicotine, for example, both impair insect development; although insects may continue to feed on the plant, their numbers are kept low and the damage is limited.

As well as defending itself from infection and predation, a plant must also reduce competition from other plants for light, water, and nutrients. Some plants secrete secondary compounds into the soil that inhibit the growth of other plants. The roots and leaves of walnut trees (*Juglans nigra*) release juglone, which harms most other plants trying to grow nearby and can even kill apple trees planted too close. Like many other secondary compounds, juglone's bioactivity has great value in herbal medicine, where it is used to kill internal parasites.[3]

Many secondary compounds are volatile. They dissipate into the air over great distances and may be detected as odors. While being eaten, plants often release volatile chemicals (such as methyl jasmonate) that are similar to compounds signaling pain in animals. They are detected by other plants as they diffuse through the air. In other words, they act as chemical warning signals. As acacia trees are grazed by giraffes, neighboring acacia trees detect the chemical warning signals and take protective measures by sending more defensive, astringent tannins to their own leaves. Giraffes, though, have learned from experience that acacia leaves start to taste unpleasant after a short while, so they move on to graze some distance away on plants that have not been warned of attack. This sort of chemical warning system is not restricted to land plants. When amphipods (small shrimplike crustaceans) graze on brown algae, the algae release water-borne chemicals that are detected by neighboring seaweed that then protects itself.[4]

Volatile secondary compounds do not just warn other plants: they can also be a cry for help. As a spider mite eats its way into the flesh of a cucumber plant, the plant releases volatile compounds that attract predators of the spider mite, which home in and eat it.[5] Many of these volatiles are terpenoids (biosynthesized from blocks of five-carbon atoms), excellent both at keeping insects away from a plant and at preventing other plants from germinating nearby. Camphene and pinene from pine and juniper trees, cineol from eucalyptus, and thymol from thyme are all volatile plant oils that are powerful antimicrobials. When plants containing volatile oils are wounded, the oils they bleed can solidify to form a protective antiseptic resin.

Other terpenoids such as sesquiterpene lactones are used in herbal medicine for their antitumor and antiulcer properties, as well as their tonic effects on the mammalian heart.[6] Taxol, a diterpene from the

Pacific yew (*Taxus brevifolia*), is active against both solid and leuke-mia-like cancers.

Some compounds are produced by a plant for the purpose of *attracting* animals. The wonderful fragrances of flowering plants evolved not to satisfy our aesthetic senses but to attract pollinators from afar. Similarly, the smell of ripening fruit attracts fruit eaters that later excrete undigested seeds some distance away, depositing them with their own starter pack of fertilizer. Smell can be far more effective than visual display, especially in dense forests or over vast distances. Elephants can detect the aroma of ripening fruit from over 20 kilometers away.

The way plants manipulate animal behavior is often quite refined. Some fruits even have a specific laxative effect so that the seeds do not spend too long in the animal's digestive tract, and some tropi-cal fruits contain sweetness-enhancing compounds that make sour substances taste sweet, thus encouraging the eating of more fruit and the distribution of more seeds. To avoid fruit's being eaten or de-stroyed too early, before the seeds are ready for dispersal, unripe fruit contains unpleasant-tasting, sometimes toxic, secondary compounds (usually tannins or alkaloids) that are gradually broken down dur-ing ripening. In the meantime, tannins act as natural preservatives (which is why they are used to preserve leather), preventing early de-composition of fruit by fungal and bacterial attack.[7]

Widely spread throughout the plant kingdom, secondary com-pounds called flavonoids are found in most aromatic plants, particu-larly fruit and vegetables. They are thought to attract pollinators, de-ter pests, and protect plants from ultraviolet radiation. Flavonoids are currently being subjected to intensive research into their ability to protect against degenerative diseases such as cancer and coronary heart disease. Salicin from willow bark is perhaps the best-known flavonoid — a natural painkiller and origin of the commercial anal-gesic aspirin.

To protect itself, then, a plant produces many different types of secondary compounds or phytochemicals (plant chemicals), each with a range of bioactive properties. The fruit of the common apple tree (*Prunus* sp.), for example, contains 153 known phytochemicals, at least 67 of which are bioactive with a broad range of medicinal actions. Another common plant, the dandelion (*Taraxacum officinalis*), contains at least 64 bioactive compounds.[8] Clearly, plants produce an array of compounds capable of influencing the health and behavior of other living organisms.

How can compounds that are designed to be harmful be *medicinal* in other contexts? Potentially an animal could benefit analogously from a plant's offensive chemicals — gaining, say, protection from bacterial infection through the plant's own antibacterial compounds. In other cases an animal could benefit from dose-related effects. Just as one paracetamol will offer temporary pain relief but two hundred will kill, so many plant secondary compounds can be toxic *or* medicinal depending on dosage. The alkaloids found in *Senecio* plants (ragwort, for instance) can cause cumulative, even fatal liver damage to grazing herbivores; yet in small doses, over a shorter time, these same alkaloids reduce the growth of cancerous tumors. Similarly, the alkaloid solanine, found in the green parts of potato tubers, can cause birth defects such as spina bifida and spontaneous abortion in animals (including humans) when eaten in high doses or over protracted periods of time. Still, a single low dose of solanine protects mice against bacterial infection.[9] Such dose-related effects appear to play a role in much of preventive medicine in the wild. I shall return to this important point later.

Not only can different dosages of secondary compounds have dramatically different effects, but the dosage in each bite can also vary tremendously. Although it is fairly easy to identify the main groups of secondary compounds in a species of plant, it is not so easy to predict how much of any particular compound will be present at any one time in any one part of a plant. As the quantity consumed is critical in distinguishing between medicine and toxin, this variation enormously complicates research in herbal medicine and animal self-medication.

In plants with two separate sexes (dioecious), male plants are generally more heavily grazed than females, suggesting that female plants may contain greater concentrations of feeding deterrents. This is certainly the case for the female Utah willow, which has much higher levels of tannins and salicortin than the male. Not surprisingly, herbalists often specify the sex of a plant in their medicinal preparations. In traditional Belizean herbal medicine, for example, the female ki bix (cow's hoof) is used for birth control, whereas the male plant is said to be better for dysentery and hemorrhaging.[10]

Different parts of plants are defended differently, depending on their value to the plant and how often they are attacked. In the boreal forests of North America, the upper branches of trees, which are beyond the reach of the tallest moose, have far fewer secondary compounds in their leaves than those lower down, which browsing moose can easily reach. Similarly, in eastern Africa all the distasteful phenols of acacia plants are in their lower branches. Seasonal changes occur too. Those same boreal trees have higher concentrations of secondary compounds in winter, when browsing by hares, moose, and rodents is most intense.[11]

Because plants can move their defensive compounds if and when necessary, different parts of the same plant will contain different medicinal compounds at different times. The type and concentration of secondary compounds in a plant will therefore be dependent not only on its inherent nature (whether it can produce a particular secondary compound) but also on its circumstances. The Mayan Indians believed that the medicinal resin of the copal tree should only be drained during a full moon. Bearing in mind the way medicinal secondary compounds can move around as environmental circumstances change, traditions of this sort no longer look ridiculous.

We should bear this fact in mind when recording what animals eat in the wild. Laboratory analyses of plant constituents can only provide a snapshot of what goes on in the plant's life. Nothing is fixed. Change is one of the few constants in nature. What is found in the laboratory represents only what is looked for at one particular moment by one particular technique. It is estimated that there are a quarter of a million species of flowering plants; only a tiny proportion (less than 15 percent) of these have been analyzed in detail, and many unexamined species go extinct every day. Although laboratory analysis can reveal certain phytochemicals, and provide useful clues as to the actions of each, the ingredients may have vastly different effects on animals eating the plant in the wild.

*

The potential plant pharmacy is vast. However, it is not the only source of natural medicines available to animals. Because many insects harvest plant secondary compounds (something we shall look at in detail in the next chapter) or make their own defensive compounds, they too are a source of potential medicines for other animals. More than seventy insect species are used in traditional Chinese medicine. Furthermore, many amphibians, reptiles, and birds take toxins from their insect diet to use in their own defense. South American arrow-poison frogs get at least some of their deadly skin toxins from the ants and beetles on which they feed. Fungi, famed for their ability to produce powerful antibiotics such as penicillin, also manufacture potential medicines for nature's pharmacy.

Secondary compounds are not the only plant chemicals that can be medicinal. Some chemicals of plant primary metabolism can affect animal health; even plant structural compounds, such as the fiber found in bark or certain grasses, can be medicinal. What is more, nature's pharmacy is not confined to living material. Even earth can be medicinal (as we shall see), and soil often contains microbial organisms that themselves secrete bioactive compounds. Since nature's pharmacy is not limited to plants, even meat-eating carnivores at the top of the food chain have access to powerful medicinal compounds.

All animals are surrounded by powerful pharmacological substances and have ample opportunity to self-medicate. The crux of the matter is whether they actively exploit this opportunity.

3

# FOOD, MEDICINE, AND SELF-MEDICATION

> Leave your drugs in the chemist's pot if you can heal your patient with food.
>
> — Hippocrates, 5th century B.C.

FOOD IS MORE than fuel. It is the substance of which we animals are made, and as such can make or break our health. Although it is accepted that lack of nutrients contributes to disease; that the consumption of excess nutrients contributes to chronic heart disease, high blood pressure, cancer, arteriosclerosis, and diabetes; and that many foods contain compounds that protect us against disease, few Westerners today adhere to Hippocrates' advice. In the past century, the pharmaceutical industry has forced a wedge between food and medicine, so that the word *medicine* is now almost synonymous with drug therapy. The industry has turned natural plant compounds into drugs, mass-producing them synthetically in far greater quantities than can be extracted from plants. It has also developed new artificial drugs that have strong, rapid, dramatic effects on the body, and, most important for the pharmaceutical industry, can be patented (unlike traditional herbal remedies). We now consider medicine very different from food: not something we do for ourselves, something we buy from experts.

*

This distinction is not so clear-cut for many non-Western (and a few Western) physicians, who still use food as medicine to both prevent and treat disease. In traditional Chinese medicine, for example, bacillary dysentery is treated with ordinary tomato plants, and bronchial asthma with steamed pumpkin.[1] Such food cures stand up well to scientific exploration. A study at Harvard Medical School found that eating more than two tomato-based meals a week reduces the risk of prostate cancer by as much as 34 percent. Nutrients therefore do more than merely keep animals alive. They can help animals stay *well*.[2]

When we look at animals, it is often difficult to distinguish between behaviors that are nutritional and those that are medicinal, because the separation is *artificial*. Food and medicine are better described on a continuum: high-energy foods, eaten primarily for the purpose of fuel intake, are at one end, and highly bioactive substances (normally considered nonnutrients), eaten primarily for medicinal purposes, are at the other. Much of what animals eat in the wild falls somewhere between these extremes, yet is still a crucial part of active health maintenance. If we are to discover how animals stay well in the wild, we need to explore their diets in some detail.

## SPECIES NICHES

Each species has energy and nutrient requirements that are specific to the role played by the species in the ecosystem, whether to eat the dung of herbivores, steal the carrion of other predators, or browse the leaves of trees. Each has evolved a different way of obtaining, processing, and digesting an optimal diet. Herbivorous sheep, for example, not only have teeth adapted for cutting and grinding up vegetation, but also enzymes for breaking down tough plant cell walls, as well as symbiotic gut microorganisms in specialized fermentation chambers, to help the digestive process. Carnivores, such as leopards, on the other hand, have great difficulty in extracting nutrients from plant matter but do have strong, sharp teeth for tearing flesh, and specialized enzymes for breaking down protein-rich meat.

If 25 million animal species exist in our world, there are probably as many different dietary habits and digestive systems. Yet animal diets have some common themes, such as the need to consume carbohydrates, proteins, fats, vitamins, minerals, and water. What an animal eats has to be right for its individual circumstances. As a general rule, sheep eat grass and leopards eat meat, but extensive flexibility is built into their behavioral repertoire to allow for changing circumstances.

The scientific study of diet is surprisingly young, and the diets we feed most captive wild animals have been developed primarily through trial and error. When gorillas were first "discovered" in the late nineteenth century and brought to Europe for display in zoos, the only way zookeepers could keep them alive was by feeding them meat, so it was assumed that meat formed an essential part of their diet in the wild. But almost a hundred years later, extensive field studies of wild gorillas found that they are almost completely vegetarian, eating a broad diet of bitter, sweet, and fibrous plants, soft fruit, and seeds, supplemented only by the occasional insect.[3]

The lack of detailed knowledge about the diets of wild animals is not surprising. It takes time, energy, and patience to habituate wild animals to human observation. It also takes extensive experience and knowledge to record everything they eat. As almost 85 percent of plants remain unexamined by science, the precise content of a plant-eater's diet is currently impossible to fathom. As diet changes throughout an animal's lifetime, only long-term studies can yield useful data. Indeed, surprises are still commonplace.

Until the late 1960s, for example, chimpanzees were assumed to be essentially vegetarian, as they had never been seen to eat animal protein other than a few insects. Then the sustained observations of Jane Goodall and her team in Gombe revealed that chimpanzees not only ate meat, but killed to get it — they hunted! Because long-term field studies are so few and far between, it is no wonder that even now, at the beginning of the twenty-first century, we still do not know the full dietary requirements of most wild animals — even those we regularly keep in captivity. Certainly, far more is known about the nutritional requirements and habits of species economically or emotionally important to us, such as farm and companion animals. But even here our knowledge is limited to an understanding of what an animal finds most acceptable in captivity, or how efficiently food can be turned into an animal "product." Questions as simple as why an animal selects a specific food item on one occasion and a different one on another remain unanswered — even unanswerable — with the data we have.

## NUTRITIONAL "WISDOM"

When laboratory scientists started to explore how animals select their diets, they quickly found that rats that are presented with a range of foods cafeteria style will select a nutritionally balanced diet. This ability, termed nutritional wisdom, can be loosely applied to the way in which wild animals manage to meet their nutritional needs from foods that are often changing in composition, availability, and location.

Although fallow deer are considered grazers, in temperate environments they generally graze only in summer. In autumn, when the grass dies back, they switch to browsing fruits such as acorns and beech mast; then in winter, when the fruits are exhausted, they browse on brambles, ivy, and holly. When the grass returns in spring, they stop browsing and start grazing again. As the food supply changes, they strive to obtain a balanced intake of nutrients and energy. This ability is not limited to mammals; insects can regulate their intake of sugars and amino acids by changing what they eat.[4]

Often animals have to compensate for changes in the nutritional quality of their food. Aye-ayes are nocturnal primates, found only in Madagascar, that eat four main types of food: seeds, nectar, fungi, and insect larvae. When the energy content of these foods drops during the cold season, aye-ayes double the amount they eat. During drought conditions, when their ordinary diet dies back, feral camels in Kenya concentrate more on evergreen shrubs and salt-rich plants such as *Salvadora persica* and *Sueda monoica*. Along with rhinos, they start eating water-rich *Euphorbia* plants — and even grow fat on them.[5]

Animals do not merely manage to select a balanced diet from highly variable foods; they constantly adapt this diet to their own changing circumstances, sometimes in *advance* of a change. Birds preparing for migration need to lay down extra reserves of fat for long journeys without food. Just before migration, several species of songbirds change their dietary preferences. Garden warblers preparing for migration switch from eating insects to focus almost solely on figs, even though insects are still available. The secondary compounds in fruits are thought to be responsible for changing the birds' metabolism so that it is easier for fat reserves to be laid down. Even though golden-mantled ground squirrels do not migrate, they do hibernate, and in preparation they change the type of fat they consume. Normally they eat fats that give off toxins as they are metabolized; before hibernation, they change to fats that will be less harmful in the months to come.[6] Selecting an adequate diet is evidently far more complex than simply obtaining the right nutrients.

## MANAGING MINERALS

The merging of nutrition and medicine is illustrated by the ways animals interact with minerals that cycle through soils, plants, and animals. Potassium is needed for a healthy nervous system, and calcium for a range of physiological processes including muscle contraction and bone formation — but it is not necessarily the amount of each that is the crucial factor. Many health problems arise when minerals are not present in the right proportions. Some minerals such as calcium phosphate play an active role in preventive medicine, providing protection against later infection with bacteria such as *Salmonella* and *Listeria*.[7] It is fascinating to observe how animals manage to find and balance their mineral requirements in the wild.

On the Serengeti Plains of Tanzania, herds of ungulates (hooved animals) distribute themselves according to the minerals in the soil and vegetation. When they have their calves, herds of wildebeests migrate from the lush grass of the north to the southern plains. There they graze at the foothills of volcanoes on grass growing on ash-rich soils, high in the calcium and phosphorus essential to lactation. During Australian droughts, when much vegetation dries up, small marsupials strip bark off trees. When magnesium licks are provided, the bark-stripping stops, suggesting that the animals have been seeking magnesium.[8]

Anyone who has kept tortoises knows how crucial it is to provide them with a piece of chalk or some other source of calcium, along with their normal food. Without the calcium, tortoises develop shell deformities and become unwell. In the wild, calcium is not readily available in the plants desert tortoises eat, because it gets impacted beneath the surface of the soil by extreme weather conditions. Desert tortoises in California travel long distances to reach suitable mining sites, where they dig down several centimeters and feed on calcium for up to 45 minutes at a time. Females mine more than males, probably because they need extra calcium for egg production.[9]

Like all mammals, rats require extra calcium during pregnancy and lactation. To meet this extra demand, they (like us) increase the amount of food they eat. They also undergo hormonal changes that ensure more absorption of calcium from the food. Even so, they may not get enough calcium, and in this case the animals actively seek out a high-calcium diet. Exactly what triggers this change in feeding behavior is not yet known.[10]

When ungulates such as moose grow antlers, they need so much calcium and phosphorus that the minerals are shunted from their bones to produce up to 400 grams of new antler tissue each day. As a result, they can suffer osteoporosis. Where soils (and hence plants) are deficient in calcium and phosphorus, deer, caribou, and moose chew on cast antlers to regain minerals. Antler chewing has been the source of some embarrassment to archaeologists, who for many years assumed that the teeth and scrape marks on fossilized antlers and bones were a sign of prehistoric hunting by early humans. In spring, when a reindeer's body reserves have been used up, yet it needs to grow antlers or produce milk for its young, it will try to acquire the needed minerals from *any* source available. Not only will it chew on cast antlers, like other ungulates, but it will eat soil from around decayed bones, lick rocks, steal salted fish, and even drink brackish water.

Strange and unusual eating habits often conceal nutritional wisdom. On an expedition in China in 1999, John Hare, director of the Wild Camel Foundation, passed by the carcass of a dead wild camel. Much to Hare's surprise, the camel he was riding stretched down and started chomping the skull in its vegetarian mouth. Thinking the camel might choke, Hare tried to retrieve the skull — but the camel persisted until it had eaten the whole thing. Osteophagy (bone eating) is also seen in other herbivores, although it is extremely rare.

One scientist working in Kenya came across a giraffe chewing and eating the bones of a gazelle carcass.[11] But even carcass-eating giraffes pale next to the shocking discoveries of Robert Furness of the University of Glasgow.

During the 1970s, Furness and his team were studying birds on the Shetland island Foula. There they found strange mutilations of ground-nesting birds: decapitated bodies of chicks, as well as live, unfledged, Arctic terns with legs and wings missing. The island has no native predators such as otters or mink, but after many years of observation the group saw a Shetland ewe carrying a tern chick in its mouth, by the head. As the researchers watched, the sheep chewed the tern's head for a few seconds, shaking and severing the body, and finally swallowing it before resuming its grazing. Later two other ewes were seen pushing a live tern chick onto its back with their noses and biting carefully at the legs, severing them. Furness found more than two hundred chicks with amputated limbs on Foula, with a resident flock of only fifty feral sheep.

Amazed, he searched for other reports of this sort of activity. He found only one short reference to the possibility that red deer on the Isle of Rhum in the Hebrides might be doing something similar. He set off for Rhum, and within his first seven days in the field he saw four instances in which deer picked up live manx shearwaters and chewed off their heads. On Rhum, though, the deer were even more adept than the sheep on Foula at extracting just what they wanted from their avian victims. Several times he saw deer nibble at the legs of live birds, and upon closer examination he discovered that the leg bones *had been completely removed while the skin and feet were still attached!* It turns out that on Rhum deer are the primary predators of manx shearwater chicks, and that eagles, ravens, and crows feed only on the remains. The soil on Rhum is thin and poor, but the grass grows lush, fertilized by shearwater droppings. What minerals the birds fail to provide with their droppings, they provide with their limbs. Other scientists have since found that, on occasion, red deer eat small rabbits; reindeer eat the eggs and chicks from birds' nests; caribou kill and eat lemmings; and white-tailed deer even eat fish.[12]

Furness has suggested that sheep and deer are targeting chicks to obtain calcium. However, since phosphorus is more critical to growth than calcium, and up to 85 percent of bone is phosphorus, this could be the element that the animals are seeking. Red deer have never been seen to eat the calcium-rich shells that cover the beaches of Rhum. Furthermore, in Australia, domesticated cattle with extreme phosphorus deficiency display an avid appetite for bones. Aberdeen Angus heifers become visibly sick when deprived of phosphorus: they lose weight and condition, their reproductive cycles are disrupted, and they show signs of significant bone depletion. In this state they seek old, weathered bones (which they prefer over fresh ones) but still avoid eating flesh, fat, or blood — and thereby manage to escape botulism. At the same time, heifers with sufficient phosphorus continue to eat grass and avoid eating bones. When a sick, mineral-deficient animal seeks and finds the specific mineral needed to remedy its illness, it is difficult not to conclude that the behavior is a form of self-medication.[13]

Then, too, drinking is not always about rehydration. It can sometimes be a way of obtaining valuable nutrients, especially minerals, and perhaps medicines. In the Cameroon, African forest elephants make their own mineral licks at springs scattered throughout the forest. They excavate with their trunks and scoop up mouthfuls of mineral-rich water.[14] In the remote jungles of Pahang, Malaysia, one hunter watched a Malay elephant searching for the minerals it needed: "With a mighty blast from his trunk, he blew all the dirty water away from the mouth of the spring and got busy . . . He was very careful, when he put his trunk down, to get the exact place where the sulphur water issued from the rocks, at the bottom of the pool, and I could see quite plainly that, sometimes, he had difficulty in satisfying himself that he really had got the right spot. He evidently liked his medicine undiluted."[15]

## SEEKING SALT

Sodium is particularly valued by all land animals, as it is lost in urine and sweat and must be continuously replenished. It is so vital to human health that, in the past, salt (sodium chloride) was used as a form of universal currency. Roman soldiers were even paid in salt, from which the term "salary" derives (Latin, *salarium*). When coins were invented, they were embossed with a hallmark (Greek for salt, *hal*) denoting their equivalent weight in salt. Even today, salt is still used as money among the nomads of Ethiopia's Danakil Plains.[16]

Herbivores are vulnerable to becoming short of sodium, as plant food in mountain and inland environments can be deficient in this mineral. So strong is their desire that herbivores will risk being killed to gain access to salt. Indeed, hunters have long used salt blocks as bait to entice their quarry out into the open, or have laid in wait for them at the edge of natural salt deposits.

In Scotland, reindeer herders used to be driven to distraction by the sound of reindeer licking and chewing at the herders' salt-preserved canvas tents. Nowadays, they get more sleep: modern tents are made of chemically treated artificial fabric. In Africa, buffalo eagerly lick salt-encrusted plants, rocks, and even other sweaty buffaloes. In Central Africa's dry season, thousands of snout butterflies drop in on forest elephants to lick salt from their skins. But sweat is not the only source of saltiness. Many animals avidly lick urine for the sodium it contains. Pet dogs can even be a bit of a nuisance in the bathroom in their quest for salty urine. Reindeer, too, can apparently be a hazard to urinating herdsmen.[17]

Where possible, herbivores seek out bogs, marshes, and rivers for aquatic plants that contain higher levels of sodium (and other minerals) than land plants. Nearer the coast, many graze on salty sea-blown grasses or make their way to beaches to feed on mineral-rich seaweed and kelp. Witness reindeer in the Cairngorms, deer, and sheep of the Scottish highlands and Greek isles. Even the carnivorous polar bears of Alaska seek out seaweed when seals are out of season.[18]

It is the *taste* of salt that is sought. Perhaps because sodium is so important, salt is one of very few nutrients for which mammals have a specific hunger. Rats have specific salt-activated taste buds that become "active" when salt levels drop in the body. In short, salt tastes better when salt levels are low, and this motivates the rats to go out and find some. Jan Schulkin feels that because salt is often found in combination with other minerals, this innate "salt hunger" provides a simple but effective method by which animals may find *all* the minerals they need.[19] Finding minerals, however, is only part of the problem.

At Isle Royale, on the border between the United States and Canada, moose have a choice of feeding on land plants, water plants, or deeper forest plants. They seem to move among the three options in a regular pattern throughout the day. In summer, a moose must "consider" how much heat it will gain or lose, how much food it can store in its ruminating stomach at any one time, how long any food item is available, and how much time it needs to spend on activities other than feeding. Moose, like all mammals, require sodium for maintenance, growth, and reproduction, but at Isle Royale the soil lacks sodium. To get enough sodium from land plants, moose would have to eat more than their stomachs could hold. So they eat as many aquatic plants (pondweed, water lilies, horsetails, and bladderworts) as they can find. These plants are richer in sodium than land plants but are available only in summer (they disappear under ice in winter). To get enough sodium to last year-round, moose have to eat *more* than they need in summer. Because aquatic plants are bulkier and contain less energy than land plants, by carefully changing what and when they eat, moose can maximize their intake of energy *and* get enough sodium to last all year round.[20] No one is suggesting that moose make conscious, dietary analyses — rather, that moose dietary behavior has been shaped by natural selection to take account of their dietary needs in the future as well as in the present.

## FINE-TUNING

Micronutrients (or trace elements) are substances such as copper, manganese, zinc, selenium, and chromium that animals need to ingest in minute quantities. Too much of any micronutrient can cause poisoning — even death — but trace amounts are vital. A copper deficiency, for example, impairs immune function in cattle, rendering them more susceptible to bacterial infection and parasite infestation, and selenium deficiency renders livestock animals susceptible to disease. Zinc deficiency in pigs, sheep, and rats leads to prolonged difficult labors, and chromium supplementation helps reduce shipping stress in cattle.

Somehow, animals are usually able to obtain the right balance of these micronutrients. One way seems to be through physiological change. Rats fed on diets with widely varying manganese and zinc concentrations still manage to maintain almost constant blood levels of both by absorbing more or less of the micronutrients from the food they eat.[21] But in cases of extreme deficiency there may also be behavioral mechanisms for obtaining micronutrients. The biologist Lyall Watson witnessed a herd of cattle crowding around a tree, eating and licking its bark — something they had not been seen to do before. It turned out that the tree had one copper nail embedded in its bark where the cattle's attention had been focused — and these cattle were diagnosed as copper deficient.[22]

It is evident that everyday feeding behavior — selecting the right diet, at the right time, and for the right conditions — is inextricably linked to health maintenance.

How animals manage to find and balance most of the nutrients they need is the subject of much research. One possibility is that animals choose foods that are immediately pleasing to their senses, and as a result of natural selection their senses lead them to the right diet — sweet is appealing because it indicates energy-rich foods, bitter tastes less pleasant because it could indicate toxicity, salt tastes good because it indicates the essential mineral sodium. This process, called hedonic feedback, requires an assumption that what an animal fancies changes according to need. Another possibility, postingestive feedback, is that animals learn the consequences of what they eat and adjust their feeding accordingly. Slugs can detect the carbohydrates and proteins they need in food, compare it with the amounts in their bodies, and change their diets appropriately. Sheep are also able to monitor the carbohydrate and protein content of the food they are eating and adjust their feeding accordingly. Individual learning also seems to play a role. Sheep, for example, develop a preference for foods that previously corrected a deficiency in phosphorous.[23]

Research on livestock species — deer, sheep, and cattle — show they use a *combination* of all these mechanisms to find an appropriate diet for their ever-changing circumstances. Apparently, animals do not need to *know* what is missing from their diets to be able to rectify the lack. Although rats normally avoid unfamiliar foods, when they have been deprived of thiamine (an essential amino acid) they will try new foods in an attempt to rectify their malaise. They are not seeking thiamine specifically, but are searching out new foods in the hope of finding something that will rectify their nutritional need. They also rapidly learn to avoid foods that are deficient in essential amino acids.[24] Together these two strategies could ensure that rats find a range of amino acids sufficient to maintain health.

## NUTRITIONAL MEDICINE

Animals clearly eat food that both prevents and cures ills. They are able to find substances that protect against future illness and, as in the case of phosphorus-deprived cattle and sheep, seek out unusual substances that remedy ill health.

Most scientists prefer to describe this sort of behavior as self-regulation or homeostatic behavior rather than medicine, because the substances involved are normally considered nutrients. The homeostatic mechanism acts rather like this: when the amount of a particular nutrient in the body drops, the shortage is detected by internal sensors that trigger a search for the nutrient. When the particular nutrient is found and eaten, the body's levels of that nutrient return to normal and the searching is inhibited. So far, the only nutrients for which mammals have been found to have a *specific* search-and-find mechanism are salt and water (with tentative suggestions of similar appetites for calcium and phosphorus in rats). Specific neuronal pathways for each nutrient are unlikely. Furthermore, homeostatic models cannot explain examples where animals search for minerals in advance of any mineral deficiency. When blue tits (*Parus caeruleus*) are nesting, the males search out empty snail shells and bring them back to the nest for the females to eat. This behavior occurs only during egg laying, when the females probably need extra minerals from the shells.[25]

It is also difficult to apply the homeostatic model to cases in which animals actively seek and consume nonnutrient toxins. According to nutritional science, animals should search for nutrients but avoid toxins; still, as we have seen, a toxin is not always a toxin. Since the 1970s it has been known that certain insects consume toxic secondary plant compounds and store them in their own bodies, and in this way gain protection from predators and pathogens. But do they harvest the toxins *in order* to protect themselves from predators, or is it an incidental side effect of feeding on some other nutrient? Can we

Male Danaine
butterflies gather
toxic pyrrolizidine
alkaloids from a
drying wound on a
*Heliotropium* plant.
*Michael Boppré*

know whether an animal is eating a particular substance for nutritional or medicinal purposes? In the 1980s, the German entomologist Michael Boppré discovered that some insects concentrate *only* on a plant's secondary compounds. Danaine butterflies search out plants containing some of the most toxic plant compounds known, pyrrolizidine alkaloids (PAs). To get these toxins, butterflies have to scratch the surface of the plants and ingest the oozing sap. If the plants are withered and dry, the butterflies secrete a dissolving fluid onto the leaf and drink the resulting toxic solution. Gathering these alkaloids is therefore a special activity, *separate* from feeding behavior. Boppré concluded that drug-eating (pharmacophagous) insects "search for certain secondary plant substances directly, take them up, and utilize them for specific purposes other than primary metabolism or merely food plants."[26]

Fitness is increased by the specific searching for and gathering of these drugs, but is this *"medicine"*? When we speak of a medicine, we are usually referring to something intentionally given to treat disease. We therefore assume that *self*-medication must also be an intentional act. This assumption is misleading. In the study of animal behavior, intentional language is often used, as it is less cumbersome than repeatedly using long-winded evolutionary explanations. We might say, for example, that males compete for females in order to produce more offspring than other males; but biologists do not really think that males *intend* in any conscious or deliberate way to produce the most offspring. Their real goal may be to win females, but the production of offspring is merely the result of competitive and sexual behavior that provides more immediate rewards. Similarly, self-medication can be the result of an animal's actions, even if not its intention. In the same way that diet selection is motivated by the relief of unpleasant sensations such as thirst or a specific hunger, so self-medication is most likely motivated by the removal of unpleasant sensations or the attainment of pleasant ones.

We may glimpse such processes when we use substances that make us feel better without being consciously aware of their role. An example currently being explored in medical research is excessive smoking by sufferers of schizophrenia (more than three times the smoking rate of the general population). High doses of nicotine apparently help relieve the worst of their symptoms, yet the sufferers are not aware of the pharmacological actions of nicotine — they merely know that they like to smoke a lot. This is not the same as when we self-medicate with herbs or antibiotics that we know will help our condition.[27]

How can we be sure if an animal we are watching in the wild is self-medicating? Evidence of preventive medication is extremely difficult to find in the field and is usually circumstantial. Curative self-medication is easier to discern, and Michael Huffman of Kyoto University has devised a set of guidelines to help field biologists. First, the animal should show signs of being ill (preferably with some quantifiable test as evidence of sickness). Second, it should seek out and consume a substance that is not part of its normal diet and preferably has no nutritional benefit. Its health should then improve (again established quantifiably by blood tests or fecal analysis) within a reasonable time scale commensurate with the known pharmacology of the substance. Laboratory analysis of the plant or substance is then needed to establish that the amount consumed contains enough active ingredients to bring about the changes seen.[28]

Not many observations can satisfy all of these criteria. First, an animal may be ill without any signs discernible to an observer. This means that much self-medicative behavior can go unnoticed — especially if the medication works! Second, the medicine may not be an *unusual* substance. As we have seen, everyday plant foods can contain powerful medicines. Perhaps the animal changes the amounts, proportions, or ratios of ordinary foods. Third, not all self-medication will be successful — unless animal self-medication is far more effective than human medicine. Finally, it can be misleading to focus on the effects of "active ingredients" on an individual's physiology. Plant medicines, made of whole plants, do not work like single isolated ingredients. They can alter metabolism in a general manner or can affect a complete hormonal system, and mixed ingredients can enhance or antagonize each other. Tiny amounts of powerful molecules can have dramatic physiological influences, whereas large amounts of another compound may be metabolized and excreted without effect. Documented cases of curative self-medication that meet Huffman's criteria therefore probably reflect only a tiny proportion of the total amount of animal self-medication taking place. However, these criteria are crucial in establishing a solid base from which a scientific exploration of animal self-medication can emerge.

By far the easiest way to distinguish between self-medication and nutrition is when animals use medicinal substances without eating them, as when they rub bioactive compounds into their fur or feathers, or use them to fumigate their nests. We shall see that wild animals use not only plants, but also soils, rocks, mineral-rich waters, sunlight, toxic insects, bark, and charcoal in ways that can only be described as medicinal.

# 4

❧ ❧ ❧ ❧

# INFORMATION FOR SURVIVAL

> To acquire knowledge, one must study; but to acquire wisdom,
> one must observe.
> — Marilyn vos Savant

OBSERVING HOW ANIMALS BEHAVE has always been essential to human survival. Cave paintings show that our early ancestors had a
firm grasp of the behavior of the animals they hunted, and people
have used animal behavior to forecast changes in the weather or
warn of impending natural disasters. The arrival of seabirds on the
shore can predict a storm at sea; and before major earthquakes small
mammals evacuate their burrows, swarms of insects gather near the
seashore, and cattle migrate to high ground. Fishermen have noticed
that before earthquakes, Japanese deep cold-water fish congregate in
the shallow warm waters in the Sea of Japan.[1] Sometimes the benefits
of watching animal behavior are immediately obvious, as when Joy
Adamson noticed her orphan cheetah stop and stare at something
unseen, which saved her from walking into the path of a poisonous
snake. At other times the benefits are apparent only with experience.
If you are ever stuck without water in the Nevada desert, watch for
desert tortoises digging shallow holes in the rocky soil. They are preparing to collect the surface runoff from an impending storm.[2]

Yet it is wrong to view animals as all wise and all knowing. Rather,
they demonstrate successful strategies for survival that have been
honed by natural selection. Many of their seeming powers of predic-

tion result from different sensory equipment: they may *hear* the subacoustic rumblings of an earthquake, or *feel* the changes in atmospheric pressure or electromagnetic field as a storm approaches.[3] The senses of other species can at times be more acute and reliable than our own, but we need a great deal of experience to discriminate between when this is so and when it is not.

Our observation of the world around us has obviously changed dramatically with the advent of sophisticated technology, and the study of animal behavior is no exception. Behavior is now commonly broken down into discrete predefined actions and the frequency of these actions is compiled on computerized event recorders. Unless a particular behavior has been defined as worth recording, it may pass unnoticed. Until recently, for example, many observations of "homosexual" mating were ignored. Then in 1999 Bruce Bagemihl wrote an exhaustive treatise showing that most animals do engage in homosexual encounters, and scientists (myself included) had to admit that for years they had ignored observations of male-male and female-female sexual activity because either it did not seem relevant or it did not fit with current theories.[4]

Another side effect of technology is that few ethologists actually go into the field to observe real animals — spending hours waiting; getting wet, cold, and hungry; and perhaps putting their lives in danger. Instead, we can observe computational models of animal interactions, or attach satellite collars to animals and observe their blips moving across a computer screen from the comfort of our desks. Already, anyone with access to the Internet can trace the movements of forest elephants across Malaysia, sea turtles across the Pacific, and wolves across North America. Many of the tracking devices are able to inform us whether the animal is active or resting, often in locations where observation would be extremely difficult. As these ingenious devices record more and more detailed aspects of behavior, ethologists may no longer feel that they need to observe wild animals in the flesh at all — and without direct observation, we risk missing much vital information.

A vivid illustration of the value of direct observation of wild animals comes from the work of Monty Roberts, the "horse whisperer."

When Roberts was only thirteen years old, he went out into the Sierra Nevada and watched feral mustangs for three years. He saw how they communicated with one another, what signals they used to convey emotions, and, most important, how they settled disputes.

Not being a trained scientist (or even an adult), he did not follow the herd in a jeep, holding a computerized event recorder and analyzing the frequency of discrete behaviors. He did not pull down the horses, attach radio collars, and plot their movements on a map. He simply rode a horse alongside the mustangs, silently observing them for hours, days, weeks, absorbing the body language, the nuances, the glances, the dynamics. From these observations he founded a whole new system of working with horses. The normal method of breaking young horses involves weeks of grueling physical and psychological battles between man and horse. Monty learned how to use horse body language to get a horse to cooperate within thirty minutes. His touch with horses has been described as unique, even mystical, yet his techniques have been adopted by other horse whisperers who have not had his direct experience with wild horses.

At age sixteen, when he demonstrated to his father how he could start a wild horse, his father beat him. It was another forty years before Monty was brave enough to show anyone else. When he first suggested that horse breakers did not need to *break* horses — that they could win cooperation by learning horse body language and *start* them instead — he was ignored or ridiculed. It was only in 1989, when the Queen of England, knowledgeable about horses herself, summoned Roberts to put on a demonstration at Windsor Castle, that other horse trainers began to take him seriously.[5]

Simply observing the world without manipulating it is undervalued nowadays, and for valid reasons. Human perception is heavily biased toward occurrences that have been relevant for survival in our own evolutionary history, and this affects the way we link associated events into cause and effect (particularly when we have inadequate information).

A classic piece of animal folklore illustrates how observations can be misinterpreted. In first-century Rome, Pliny the Elder saw that when snakes were blind they rubbed their eyes against fennel and restored their eyesight. He concluded that fennel could cure eye problems. Other herbalists subsequently reported similar observations, and as recently as 1987 Edward E. Shook wrote: "When the snake casts his skin he goes blind and, in that hopeless condition, he has often been seen to crawl to an old fallen tree or a mossy bank. He rubs his sightless orbs in the moss and eats the rich juice, and with that one herb (moss) he restores both his skin and his sight."[6]

Although these observations may be accurate, the interpretations are wrong. Snakes regularly and periodically shed their skin. A few days before shedding, snakes are effectively blind because fluid builds up between the new and old layers of skin. In this state, the eyes look clouded and grayish blue. To shed the old skin, the snake must start a break, and the easiest place to do so is along the edge of the mouth or jaw. Thus, shedding snakes will rub their heads on plants, mossy banks, or other suitable objects. Once the skin has been successfully removed, the eyes become clear and the snake's "blindness" is cured — not by medicinal plants, but by *mechanical* removal of the skin.

Valid observations followed by incorrect interpretations are common in folklore about animal self-medication. In the Middle Ages, English physicians observed that mammals always lick their wounds and that healing seemed to be hastened thereby. They concluded that the tongues of mammals had some amazing healing property. Accordingly, extracted puppy tongues were prescribed and used as wound dressings. Today we know that it is saliva, not the tongue, that contains antiseptic and wound-closure agents.

We need scientific methodology to steer us through the rocky terrain of sensory, association, and interpretation errors, and to counterbalance our predilection for superstition, any experimentation should be preceded by a period of unbiased, open-minded, and direct observation. A hypothesis can then be formed and experiments designed to test the hypothesis. Other scientists can repeat the experiment to verify the validity of the research. Following the scientific method can save us from jumping to erroneous conclusions based on false associations, or on the behavior of one peculiar animal, or on the whims of a scientist's personal agenda.

To learn anything useful, it is vital to keep going back to direct observation for fresh ideas. As we enter the twenty-first century still carrying heavy burdens of disease and chronic ill health, scientific methodology must be applied to the study of how animals stay healthy in the wild, and this process must start with observation.

Since ancient times, humans have learned about potential plant medicines by observing wild animals. Chinese folklore records that in the Han Dynasty (206 B.C.–A.D. 220) a general called Ma-Wu and his defeated army retreated to a remote region where food and water were scarce. Many horses and soldiers died, and those that survived were weak and sick, nearly all of them excreting blood in their urine. One groom, noticing that his three horses were fine, watched closely what they ate and saw that they were consuming large amounts of a small plantain plant. He boiled some for himself, and a few days later the blood in his own urine had disappeared. He gave the herbal drink to other men and horses and they too were cured. When the general asked where he had found the plants, the groom replied cryptically, "I found them before the cart." So the plant (*Plantago asiatica*) is now known as Plant-Before-Cart. It has been found to contain iridoids, flavonoids, and tannins (among other ingredients) and is anti-inflammatory, diuretic, and antimicrobial.[7]

In seventeenth-century England, physicians considered observation of animals a reputable method of learning about new medicines. Dr. Jacquinto, physician to Queen Anne (consort to James I), regularly went into the marshes of Essex, where local people put their sheep to cure them of a disorder known as "the Rott." There he observed the plants they ate and thereby discovered a useful mixture of herbs to treat humans with consumption.[8]

Native American Indians have a long history of learning medicine from observation of animals, of the bear in particular. Since the bear slept all winter and reemerged each spring, people saw him as an animal capable of regeneration and healing — coming back from near death each year. An omnivore like man, the bear eats roots, leaves, berries, and fish — much as the Native Americans do. So when people saw a bear using a plant that appeared to heal an ailment, they suspected it would be good for them too. The bear was so revered that only the most advanced healers in Native American society were honored with the name Bear Medicine Men.[9]

There are many more recent examples of humans learning medicine from animals. One involves a traditional herbalist, Kalunde, of the WaTongwe people of Western Tanzania. Kalunde rescued an orphaned porcupine after its mother had been killed. When the young porcupine became ill, suffering diarrhea, lethargy, and abdominal bloating, it went into the forest and dug up and ate the root of a plant called *mulengele*. The porcupine soon recovered, and Kalunde subsequently found that the plant was useful for treating internal parasites in his human patients.[10] Similarly, after Creole traditional herbalist Benito Reyes, from Pedregal, Venezuela, observed deer, peccaries, and other animals chewing and scraping the hard astringent seed cases of Cabalonga (*Nectandra pichurim*), he found it efficacious in treating his human patients who suffered from diarrhea, colic, sore stomach, and nervous disorders.[11] Herbalist Juliette de Baïracli Levy writes, "I have learnt much herbal medicine from the wild creatures, noting the herbs that they select for their food or medicine." And the contemporary herbalist Maurice Mességué was taught by his father: "You learn by looking — see, these animals know more about it than we do. They know the plants and the grasses, which are good for them and which are bad, they know what to eat and how to take care of themselves."[12]

Despite the value practitioners of traditional medicine put on animal observation, scientists have only recently begun to study animal behavior. For it was in the 1960s that Niko Tinbergen devised the four main questions of ethology: What is the immediate *cause* of the behavior? What is the *purpose* of the behavior? How did the behavior *evolve?* and How did the behavior *develop* in this individual?

The first behavioral studies of elephants and chimpanzees started in the 1960s, of wolves in the 1970s, of jaguars in the 1980s. Most species still have not been studied at all. Small wonder, then, that animal behavior has not played much of a role in Western medicine and health. The most important reason, however, is that in medical research animals are regarded not as teachers, but merely as physiological "models" of humans. Consequently they are subjected to a range of often painful, and ultimately fatal, medical procedures in investigations of diseases, drugs, and surgical interventions.[13]

Noninvasive systematic observation of healthy animals is unheard of in medical research. New cures are discovered through deduction and laboratory-based experimentation, and are certainly not learned in any *receptive* way by observing other animals. In veterinary care, as in human medicine, huge sums of money are invested in fighting disease with gene therapy, expensive drugs, surgery, and transplantation of organs, while virtually nothing is spent on looking at how healthy individuals successfully keep disease at bay. A cynical view might be that there is little profit to be made in "health," but much profit in treating recurring disease. A more generous explanation is that we are so busy fighting sickness and disease in ourselves and other animals that there is little time left for preventive medicine.

Now scientists have begun to study the actions animals take to deal with health hazards. Because it is impossible to follow all the stages of scientific method in the wild, we often have to rely on anecdotal and circumstantial information, which is vulnerable to claims of being unproven or untested. As geneticist Richard Lewontin explains, "The experimentalists look down on the observers as merely telling uncheckable just-so-stories."[14] Yet long-term field studies of wild animals have taught us lessons that laboratory experiments never could. The way animals interact with one another, and their environment, simply cannot be adequately studied in a laboratory setting.

Jane Goodall, who has spent forty years studying the behavior of wild chimpanzees in Africa, was criticized early in her career for reporting her observations of single behaviors performed by individual animals, especially since she insisted on referring to the apes as "he" and "she" instead of "it." Science has moved on since then, but it is still hard to have a single observation published. Goodall maintains that "in the wild, a single observation may prove of utmost significance, providing a clue to some hitherto puzzling aspect of behavior."[15] It was one such observation that alerted the world to the possibility that chimpanzees make and use tools. Since then, many more tool-making chimpanzees have been seen by a number of different ethologists, but the idea that tool making should be *looked for* as a distinct possibility was based on Jane Goodall's observation of a single behavior by a single male.

Another revelation has come from long-term studies of wild chimpanzees: since the 1980s, ethologists have been reporting anecdotal observations of chimpanzees apparently using plant medicines to prevent and cure sickness (this will be described in detail in Chap-

Michael Huffman and Mohamedi Seifu explore chimpanzee self-medication in the Mahale Mountains of Tanzania. *Michael A. Huffman*

ter 9). The first published account, in 1989, described the behavior of one sick female chimpanzee that sought out a rarely eaten plant and then recovered.[16] A flurry of media excitement and popular interest resulted, although of course one observation of one individual was not enough to establish that wild chimpanzees routinely medicate themselves. It merely pointed out the possibilities for further research.

The laboratory-based neurophysiologist Robert Sapolsky was swift to respond to this report in *The Sciences*. "The instances they postulate as animal self-medication have not yet been proved, nor have the explanatory mechanisms underlying any putative cases." He added, "Once it is proved, I'm as taken with a romantic fact as the next person." There is an understandable aversion to the vision of wild animals as wise beings who knowingly heal themselves with herbs from the woods and glades. Later, in a different article, Sapolsky wrote: "They [animals] sense impending earthquakes, intuit human emotions, predict downward trends in the stock market. It is a seamless kind of wisdom that reflects a profound equilibrium with the environment. Much of this is nonsense. Animals sometimes do indeed perform amazing feats of intelligence. But the idea of *animaux savants* is soggy with romanticism."[17]

The idea of animal self-medication touches a nerve mainly, I suspect, because of its historical links to herbalism and folklore — and ultimately to the prescientific era of witchcraft and superstition. However, the scientific study of animal self-medication need have no romantic undertones. The ability of animals to find relief from unpleasant sensations is at the root of much of their behavior and is highly adaptive. Certainly, we should steer clear of anthropomorphizing (of putting human feelings and attributes onto animals), but we should equally steer clear of doing the opposite (of *denying* recognizable attributes in animals because they resemble human traits). Indeed, despite his earlier warnings, Sapolsky now accepts that progress has been made in animal self-medication research and that more work is needed on this fascinating subject. Bearing in mind that it is early in this new science, let us now look through a wide-angled lens at how animals in the wild deal with ubiquitous health hazards.

# PART II
# HEALTH HAZARDS

# 5

✄ ✄ ✄ ✄

# POISONS

A wild animal living free never poisons itself, for it knows what foods to choose. This is an instinct animals lose when they are domesticated.

— Maurice Mességué, 1991

THE MANAGERS of Micke Grove Zoo in California spent a great deal of time and money designing and building a wonderful outdoor island compound for their black and white ruffed lemurs. Natives of Madagascar, these primitive primates forage in the tropical forest for sweet fruits, insects, and occasionally small birds and mammals. After a year of preparation, the small group of zoo lemurs were finally released into their open space, amid much anticipation and excitement. Yet within twelve days one was dead, a second moribund, and a third clearly sick. The hastily called veterinarian found the survivors unresponsive — their pupils dilated, their coordination disrupted. Blood tests could locate no obvious infection. Soon a second lemur was dead. Distraught keepers searched for clues. Had some unwitting or malicious visitor thrown poisoned food into the compound? Sadly, the answer turned out to be much simpler — and in some ways harder to bear. The lemurs, unfamiliar with their new surroundings, had sampled all the plants growing there, including several specimens of hairy nightshade (*Solanum sarrachoides*) of which the keepers were unaware. Its toxic alkaloids had poisoned them.[1]

*

Herbalist Maurice Mességué overstates his point by claiming that wild animals *never* poison themselves, but considering that an estimated 40 percent of plants contain some kind of defensive chemical, animals in the wild do appear amazingly adept at avoiding the worst of their effects.[2] And certainly he is right in asserting that domesticated animals more readily succumb to plant poisoning. Not surprisingly, natural selection hones the skills needed to avoid and deal with poisoning.

Wild rats are much better than domestic rats at detecting rodent poisons, at redirecting their feeding to safe foods, and at retaining a long-term aversion to foods that have poisoned them. Domesticated cattle and horses are particularly vulnerable to being poisoned by plants of the genus *Senecio,* such as ragwort, which contain pyrrolizidine alkaloids that cause cumulative liver damage. Between 1988 and 1992, unusual weather conditions in Queensland, Australia, favored *Senecio* plants over other plants. As a result, domesticated cattle died by the hundreds. Not only were the cattle unable to detect the toxicity of *Senecio* and thereby avoid eating it, but they were also unable to detoxify it once eaten. Wild species that have evolved with *Senecio* in their environments are much better able to deal with it. Wild black-tailed deer can consume up to 24 percent of their body weight in *Senecio* for over a month without being poisoned, and the wild mountain viscacha, a large rodent, eats *Senecio* as a major food item. Clearly, the more we know about how each species naturally avoids poisoning, the better we can manage their health.[3]

In addition to the physiological adaptations specific to each species for dealing with toxins — detection mechanisms, specially adapted biochemical pathways, or detoxifying microorganisms — each has different behavioral strategies as well.

Herbivores, with their evolutionary history of dealing with plant defensive chemicals, are generally better adapted for dealing with plant poisons than omnivores; and these, in turn, are more able than carnivores. Carnivorous alligators are easily killed by strychnine (an alkaloid from *Strychnos* spp.), while herbivorous fruit bats are able to consume high concentrations with impunity. Wild herbivores are very adept at discriminating toxic from less toxic plants, even within the same species. Bracken fern, for example, comes in two forms — only one of which contains universally toxic cyanide glycosides. Both red deer and sheep avoid fronds and rhizomes of the bracken that contain the cyanides but readily consume those that do not. Voles are similarly able to discriminate between clovers that contain cyanide and clovers that do not. Unsurprisingly, they prefer to eat the clovers with no cyanide, but if forced to eat the cyanide types, they reduce the amount eaten and store most of it.[4]

The bitterness of toxic alkaloids and saponins, and the astringency of tannins, can warn of a plant's ill effects. But it is unlikely that animals have specific sensory receptors for every potential toxin. It is more probable that when an animal regularly encounters a chemical that has significance to its survival, selective pressure evolves specific sensory capabilities to recognize that chemical. In other words, animals evolve ways of detecting only those toxins *normally found in their environment*.[5] This process explains why so many animals, like the lemurs above, are poisoned in unfamiliar surroundings. They simply cannot detect and thereby avoid the toxins.

Toxin-detection ability can be very highly developed. I have lost count of the times I have tried, unsuccessfully, to hide a bitter worming pill in my cat's dinner. However carefully I hide the lentil-sized pill deep within the chunk of attractive meat and biscuits, when the cat is finished with her meal, the pill remains alone in the bottom of the dish. The unnaturally strong toxins intended to kill the worms have been rejected.

Herbivores have at least two broad options for avoiding plant toxins. One is to specialize, to put all of their resources into dealing with a limited range of toxins. The other is to generalize, to dilute the toxin load by taking in smaller amounts of a greater range of toxins. The maned sloth lives in trees, eating a diet of leaves — but from a limited range of about ten uncommon plant species out of the huge number available in its neotropical environment. Furthermore, from those ten species it selects only young leaves, which contain less tannin. In this way the sloth, which owing to its slow metabolism has a limited ability to detoxify plant defensive compounds, limits the range *and* the amount of toxins with which it must deal.[6] The mountain gorilla, on the other hand, will eat "anything within arm's length," according to biologist John Berry. Generalist feeders have a strong drive to eat a varied diet. If fed only a single food item for a full day, a generalist grasshopper will try to eat almost *anything* (even nonfood items!) in a bid to vary its diet.

## EACH TO ITS OWN

Individual animals can also learn what is poisonous. One way to do so is by watching what others eat. But what is safe for one species may not be safe for another. In the desert of Namibia grows a plant, *Euphorbia virosa,* that has ferocious, sharp, upward-pointing spines. If that is not enough to deter a hungry herbivore, it also oozes stinking, white, toxic latex, which blisters the skin on contact. Locals call the plant Gifboom ("poison tree") and use the latex as an ingredient in an arrow poison for hunting small mammals. Not surprisingly, this plant is avoided by nearly all animals — except black rhinos. With *no* ill effects, these ancient pachyderms can devour a poison tree from top to bottom, not only eating the entire plant but getting smothered with the toxic latex. They even seem to thrive on the tree during drought conditions.[7] So if an animal is to learn what is safe by watching others, it had better watch a member of its own species. Survival schools advise their students not to assume that food is safe just because they see an animal — even a primate — eating it. They point out, though, that the opposite strategy of avoiding food that other animals avoid should be respected.

Rats are especially proficient at learning about poisons from other rats. Normally cautious about unfamiliar foods, rats increase their intake of an item fourfold if they become aware of other rats eating it with impunity. Interestingly, they do not need to actually *see* other rats eat the new food to learn it is safe; just smelling another rat is enough to indicate the new food's safety. Rats also learn from the mistakes of others. In laboratory experiments, naive rats avoid ingesting toxins already identified by others in their social group — again without having to see the other rats suffer. The mechanism used to transfer this aversion between them is still being researched.[8] Mongolian gerbils are even more discerning; they only trust their observations of relatives or close associates. What strangers choose to eat is not copied. This is a very sensible strategy, as close relatives share a similar ability to deal with toxins.

Certainly much of an individual's familiarity with its food is gained through experience. There are not simply safe plants and unsafe plants, once learned and never forgotten. There are plants that are safe in moderation, but lethal in excess; safe at certain times of the year, but not at others; safe when first eaten, but not after a protracted period of feeding; safe only after processing or preparation; and safe when eaten in combination with something else, but not alone. To assimilate all of this information requires some kind of feedback on the consequences of what has been eaten. After eating a food item that makes it sick, an animal will associate that food with the unpleasant sensation and avoid it in the future. Most of us know only too well how the taste, smell, or even sight of food that has made us sick can bring back associated feelings of nausea and disgust. This is known as a conditioned aversion, and rats are experts in forming them. The intensity of the lesson is quick and absolute, taking effect after only one exposure. And a good thing it is, for selection will not favor those individuals that need several poisonings before rejecting a food.

Mammals have an opportunity to learn the taste and smell of safe foods while in the uterus, and later from their mother's milk, as well as by sampling what she is eating. Rats develop aversions to unsafe odors, and preferences for safe tastes, while still in the uterus. Infant sheep and goats learn from their mothers and peers what is and what is not appropriate to eat, and this distinction plays a crucial role in the formation of food choices later in life.[9] Baby elephants determine what is safe by taking food out of their mothers' mouths and tasting it for themselves.

Once a suitable amount of sampling and learning has occurred, one way to eat safely is to consume *only* familiar foods, and to be careful when consuming new foods. Many species do show intense neophobia — fear of the unfamiliar — and the success of this strategy is the bane of pest controllers around the world. Rats are the most famous neophobics, sampling small amounts of any new food, then waiting, and, if they fail to become sick, returning to eat the remainder. If they do get sick, it is ordinarily only a *little* sick! People too are neophobic, at least as adults; children go through an exploratory stage when they will try almost anything, perhaps to familiarize themselves with their options. This sampling period accounts for many childhood poisonings.[10]

Sampling little and often can allow detoxification enzymes to build up in the body ("pump-priming"), making food plants effectively less and less toxic. For goats, early experience of plant toxins increases the amount they can consume as adults. Sheep gradually improve in their ability to detoxify cyanides in clover. When first exposed to clover, sheep can be killed by only 2.4 milligrams per kilogram of body weight, whereas after tolerance has built up, they can eat as much as 50 milligrams per kilogram of body weight. This tolerance has its limitations: when the amount of cyanide becomes excessive, the sheep avoid clover completely for a while, although later they will again feed on it. A similar cyclic appetite for a toxic but otherwise nutritious plant, larkspur (*Delphinium* sp.), is seen in cattle. This strategy keeps ingestion of plant toxins below lethal levels, at the same time allowing animals to obtain useful nutrients.[11]

One strange aspect of toxicity, known since the late nineteenth century, is called "hormesis" (from the Greek word meaning "to excite"). Hormetic effects are seen when exposure to a very small amount of a toxin produces a general stimulation (increased growth, fecundity, longevity, and decreased disease incidence). Although the exact mechanism of hormesis is not yet known, the best explanation seems to be that tiny amounts of toxins stimulate an overcompensation for a disruption in homeostasis.[12] This process could be extremely important in wild health, but remains largely unexplored.

However well adapted their detection and avoidance skills are, animals cannot always escape being poisoned. As veterinarian Murray Fowler explains, "Usually, when wild animals die from the effects of poisonous plants, an ecological balance has been disrupted." Extreme hunger can force them to eat plants they would normally avoid, and drought can increase the concentration of secondary compounds in plants. Sika deer are poisoned by essential oils when drought forces them to eat needles, bark, and roots of jack pine trees, for the strong antibacterial action of these oils disrupts the microbial content of the deer's rumen.[13]

Confinement can also impair an animal's ability to avoid poisonous plants. Before the mid-1960s, wild pronghorns coped with drought by migrating from the Marfa Flats of Texas to nearby hills, where browsing sustained them. When the vast plains were divided into separate ranches by barbed-wire fences, 60 percent of the prong-horns died as they were forced to eat the toxic tar brush (*Fluorensia cernua*) they normally avoided. Simply being watched can also in-crease the chances that an animal in the wild will be poisoned. Ex-periments show that the presence of a predator can distract salmon, causing them to eat inedible substances. Social position can also af-fect an animal's ability to avoid low-quality food. Social status fre-quently confers priority of access to resources, and this may include access to safer foods.[14]

It can also be difficult for wild animals to avoid being poisoned when a plant that is normally safe suddenly or unpredictably be-comes unsafe. The bongo, a large forest antelope of East Africa, is oc-casionally poisoned by the setyot vine (*Mimulopis solmssi*) when it flowers every seventh year. Similar cyclic poisonings are caused by acorns in Europe. Horses gorge on carbohydrate-rich acorns despite their high tannin content, but oak trees have their own agenda. Every few years they produce a bumper crop of acorns that floods the acorn-herbivore market, allowing more acorns to remain and grow into oak trees. In bumper years, many feral ponies in the New Forest, United Kingdom, die from tannin overdose caused by eating too many acorns. (Most years, there are no such deaths.)

To obtain adequate nutrition from plants, an animal must famil-iarize itself with the plants in its environment; yet often nutrients and toxins come together in the same plant, and the animal then needs to engage in food manipulation.

## STRATEGIES FOR MAKING FOOD SAFE TO EAT

Toxins are not always toxins: when combined with other substances, they can be safe to eat — and some animals combine foods in order to obtain nutrients safely. Tannins and saponins, for example, are harmful to mice when eaten in separate foodstuffs or when mixed in the wrong proportions, but if the mice are allowed to choose for themselves, they select a combination of the two that nullifies toxic-ity. In the right ratio, tannins and saponins bind together in the intestine, preventing their absorption into the blood.[15] Exactly how mice manage to achieve this balance remains a mystery. But they are not alone in their ability to find food combinations that reduce toxicity. Experiments show that sheep are able to avoid the worst effects of high-tannin foodstuffs if they are provided with polyethylene glycol (PEG), which binds tannins. The more tannins they eat, the more PEG they consume and thereby simultaneously increase their intake of nutrients and decrease malaise.[16] The way tannins and saponins react in the test tube suggests their combination may also reduce the likelihood of bloat, a painful digestive disorder characterized by an accumulation of gas especially in the first two compartments of a ruminant's stomach. Indeed, observations of wild leaf-eating colobine monkeys suggest that they select leaves especially high in tannins to combat bloat. Tannins are also known to inhibit the biochemical processes of cyanide poisoning, so they could be of great value if eaten in combination with cyanogenic plants.[17]

Timothy Johns argues convincingly in *The Origins of Human Diet and Medicine* that all modern cooking originated in the need to remove toxins from food and at the same time gain adequate nutrition. From simple peeling, leaching, boiling, and baking to complex fermentation processing, the complex array of human cooking is often a form of detoxication.[18]

We can see signs of food processing among other animals. In that secondary compounds are distributed unevenly throughout a plant, some parts of the plant may be safer to eat than others. Rhubarb is a common example: the stalk is edible, but the leaves contain toxic oxalates. Animals have to learn not only what species are poisonous but which parts of those species. Mantled howler monkeys in Costa Rica appear wasteful, often eating only a small part of a leaf or fruit and littering the forest floor with discarded food — but the parts they discard are toxic. Capuchin monkeys have a particularly tricky time with their food, as items in their diet vary greatly in toxicity from one moment to the next. Capuchins therefore have to assess the toxicity of each individual food item each time it is eaten. In an experiment on captive capuchins, their regular food items were sprinkled with pepper. The monkeys licked and sniffed the foods, then attempted to remove the unpleasant pepper by rubbing, wiping, and even washing the food in water. Similar strategies could come in handy when dealing with poisonous frogs, butterflies, and caterpillars, which store protective toxins on the surface of their skin.[19]

Because food toxicity changes over time, some animals use a strategy that relies on natural degradation. In the highlands of North America live small, rabbitlike, herbivorous mammals called pikas. These animals carefully manipulate the chemical content of their plant food in such a way that they avoid eating toxins *and* utilize them at the same time. In summer, pikas have two dietary strategies: plants with low phenol levels are eaten straightaway, but plants with high phenol levels are collected and stored for consumption during the long alpine winter. The pikas' favorite plant for long-term storage is *Acomastylis rossii,* a type of alpine oat high in phenols such as tannins. They are highly astringent, and pikas avoid them when fresh. But of eleven common species found at the same site, this plant is the *only one* that inhibits the growth of bacteria. The phenols gradually break down over the months, making the plant less and less astringent, and as the plant becomes more palatable, the pikas start eating their preserved winter larder. Pikas, then, are caching an unpalatable food that will not only become more palatable as the winter progresses but will preserve itself and other stored plants from spoilage.[20] The strategy of caching (and thereby degrading) toxic food may be more common than previously supposed — remember those voles that ate the nontoxic clovers but "stored" the toxic clovers?

Another way to degrade toxic secondary compounds in food is to co-opt another organism to do it for you. Central and South American leaf cutter ants actively cultivate a fungus that breaks down toxic secondary compounds in leaves, providing the ants with better access to the nutrients they need. The fungus, in return, gets a constant supply of food from the ants. Many other mutually beneficial feeding relationships may involve food detoxication.

## DEALING WITH POISONING

When mammals have eaten something extremely toxic, the signs are evident: they stagger around, clearly uncoordinated; they vomit or lie moribund; they shiver and shake. Moderate poisoning can go unnoticed, however, with only a slight reduction in activity (especially feeding), subtle signs of discomfort, slow pupil reflexes, a slight drop in body temperature, and diarrhea.

Few of us reach adulthood without experiencing the natural responses to poisoning (vomiting and diarrhea) that rapidly expel toxins from the body. Not all species have these responses, but those that do may sometimes enhance natural expulsion by consuming substances that hasten it. The gum of *Sterculia urens* fed upon by hanuman langurs in India is a bulk-forming laxative (prescribed in vast amounts as "Normacol" in British hospitals, to combat constipation). Whether this helps them deal with their strychnine-rich diet is as yet unknown.[21] Folklore is rich with examples of animals consuming emetics to induce the vomit reflex, the best known being the consumption of grass by cats.

Another way in which animals deal with poisons is by eating earth. Geophagy — the consumption of soil, ground-up rock, termite-mound earth, clay, and dirt — is extremely common in the animal kingdom. Mammals, birds, reptiles, and invertebrates have been observed to eat dirt on every continent except Antarctica. (This exception is perhaps not surprising, considering the shortage of exposed unfrozen earth.) Sometimes earth is eaten from specific sites over many generations, creating huge areas called licks that are completely bare and devoid of vegetation. Other times, animals take opportunistic mouthfuls of earth from soils newly exposed by roadwork or landslides.

There is evidence that humans have been eating earth for at least forty thousand years. During the nineteenth century explorers, anthropologists, and colonial physicians reported that native people in many tropical countries regularly ate earth. David Livingstone commented in 1870 that the Zanzibar locals suffered "the disease of clay or earth eating." The habit is still found among many contemporary indigenous peoples, including the Aboriginals of Australia and the traditional peoples of East Africa and China, as well as in the West. Often it is more common among pregnant women.[22]

Why should we and so many other animals eat dirt? People who do so often say that it "feels good" or is "tasty and sweet." Commonly, they think it is good for them without being able to tell why. Historically, the explanation was that animals ate earth in order to gain minerals, such as salt (sodium chloride), lime (calcium carbonate), copper, iron, and zinc. As we saw in the last chapter, wild animals do seek minerals from natural deposits. But sometimes this need for minerals is directly related to the need to detoxify foodstuffs.

Geophagy is far more common in animals that rely predominantly on plant food. In South America herbivorous monkeys, tapirs, peccaries, guans, pacas, and parrots regularly eat the soil of termite nests and from special licks in the forest, while more omnivorous species are rarely seen to do so. Not only is geophagy more common among plant eaters, it is also more common in the tropics. This suggests that earth eating may have something to do with the need to detoxify secondary compounds that are more prevalent in a tropical plant diet. In particular, it may relate to a need for sodium. That element is essential for all metabolic processes, including detoxication. Herbivores have a strong need for sodium not only because they may have mineral-poor plant diets, but also because they lose sodium when detoxifying and metabolizing plant secondary compounds. The extra detoxification going on in the tropical herbivore requires extra sodium, which could be provided by eating sodium-rich soils.[23]

Circumstantial support for this hypothesis occurs in observations of mountain hares in Sweden. Wildlife managers who put out salt licks for wild ungulates noticed that mountain hares used them most in seasons when dietary plant toxins were highest. Another species that appears to use sodium to detoxify its diet is the beautiful tropical great blue turaco. This bird survives on a diet rich in fruit and leaves that other animals find too toxic, and it has always been a mystery that the turaco can tolerate such leafy poisons. Recently scientists have determined that it also eats at least two species of sodium-rich aquatic plants. It is the only frugivorous bird documented to eat aquatic plants, and it is probably doing so to obtain the extra sodium needed to detoxify the secondary compounds in its diet.[24]

Elephants have a massive requirement for sodium. In a remarkable example of geophagy, a cave 2,400 meters up the side of Mount Elgon (an inactive volcano in western Kenya) has been mined by generations of elephants. It is estimated that they have carved out five million liters of rock over the last two million years. Getting to the cave is risky and arduous, with crevasses and concealed predators, yet elephants and other animals regularly make the treacherous journey. The skeletons of those that failed to make it attest to the danger.

Why take the risk? As the rains come to an end and plant growth is at its most luxuriant, the elephants start to make nightly visits to the caves. In single file they make their way through the caves, using their sensitive trunks to feel for obstacles, crevasses, and one another. They crawl through low tunnels on their knees and tiptoe carefully around large underground pools. Eventually they find the rocks they are looking for, dig out lumps with their tusks, then grind the lumps with their huge molars and eat for hours. But the animals are after more than just the rocks. Another major attraction of the caves is the mineral-rich water in the pools that attract bushbucks, buffaloes, birds, waterbuck, duikers, baboons, and monkeys, as well as elephants.

The biologist Ian Redmond found that the rocks of these caves have one hundred times more sodium than the plants normally eaten by elephants, and calcium and magnesium are present in high quantities too. As all the animals he has seen in the caves are herbivorous, he is convinced they are looking for the sodium lacking in their plant-based diet. However, the sodium in these rocks is sodium sulfate, termed "Glauber's salt" by pharmacists and used traditionally as a powerful laxative. Redmond dismisses the idea that elephants could be benefiting from the laxative effects of sodium sulfate since the elephants do not appear to have constipation. But then, if they regularly consume laxatives for several hours each night, this is perhaps not surprising! Their increased use of these mines during times of luxuriant plant growth may yet show us why elephants risk their lives in these underground caves.[25]

## CLAY AS MEDICINE

The need for sodium is by no means a universal explanation for geophagy. It sometimes occurs in locations where the soil is not rich in minerals, sometimes even with *lower* levels of minerals than surrounding topsoil. In these instances geophagy is obviously not about getting minerals. In the tropical forests of western Uganda live large colobus monkeys that spend most of their time eating leaves high up in the trees. In certain locations they come down to the ground, at great risk to themselves, to wade into shallow pools in the forest and eat mineral-rich aquatic plants. Yet they also dig out and eat clay from riverbanks. Their need for sodium (and several other minerals) is met by the aquatic plants, so why eat clay?[26]

It is also well known that humans who eat soil prefer clay to other types. In 1895 Alexander von Humboldt reported that members of the Ottomac tribe, along the Orinoco Valley in South America, regularly consumed clay in large quantities — sometimes up to half a kilogram at a time![27] In most later research, the clay content of salt licks or soils eaten by animals was not recorded — probably because clay is an inert substance, has no nutritional value, and was therefore assumed to be an irrelevant substrate. Current research, though, shows that clay is often the prime reason for the consumption of soil; its amazing medicinal properties are just beginning to be appreciated by medical science.

Clays are used commercially and industrially to bind toxins. It is the specific structure of clay, in which two or more mineral-oxide layers lie in parallel, which makes it bind so effectively with other molecules. Different types of clays, though, have slightly different structures and properties: some are better at a*d*sorbing (adding or binding other molecules *to* their own structure), while others are better at a*b*sorbing (taking a molecule *into* their own structure); some expand on absorption, others do not. In the body, toxins are deactivated by binding to clay particles, which are then excreted in the feces. Clays can bind mycotoxins (fungal toxins), endotoxins (internal toxins), man-made toxic chemicals, and bacteria; they also protect the gut lining, acting as an antacid and absorbing excess fluids, thereby curbing diarrhea. In short, clay is an extremely useful medicine.

Clay has long been used as a detoxifying substance in traditional medicine as well as in food preparation. In both ancient and contemporary cultures, clays are mixed with foods containing toxins to make them edible. Native peoples have long known to mix clay with tannin-rich acorns before making flour for bread. The Indians of the American Southwest combine clay with wild potatoes to remove the toxic alkaloids, and the Aborigines of Australia use clay to remove the bitter taste of alkaloids from roots of the *Haemodorum* species.

The benefits of clay to animal health too have been known for some time. Additions of bentonite clay improve food intake, feed-conversion efficiency, and absorption patterns in domestic cattle by 10–20 percent. Clay-fed cattle suffer less diarrhea and fewer gas-trointestinal ailments than other cattle. (In addition, veterinarians find clay an effective antacid.) In Venezuela, free-ranging cattle help themselves to clay by digging out and licking at subsoils.[28]

High in the Virunga Mountains of Rwanda, the last few hundred mountain gorillas continue to mine yellow volcanic rock from the slopes of Mount Visoke, as they have done for generations. From the size of the caverns they have carved out under the roots of trees, it is evident that these vegetarian apes treasure this dirt. After loosening small pieces of rock with their teeth, they take small lumps in their powerful leathery hands and grind them to a fine powder before eating.

Since George Schaller first documented gorillas mining volcanic rock on nearby Mount Mikeno in 1963, several other field workers have observed similar behavior. Dian Fossey reported that gorillas were far more likely to mine rock in the dry season when their diet changed dramatically to bamboo, *Lobelia,* and *Senecio* plants — all containing more toxic plant secondary compounds than their usual diet. Along with this change in diet came a synchronous increase in diarrhea (a natural response to rid the body of toxins). This extra loss of fluid, in the dry season, could potentially be a serious health problem for the gorillas. Fossey suggested that the mining and processing of the fine volcanic dust was a response to this seasonal change in diet.[29]

It turns out that although the rock the gorillas mine is low in salts, it is relatively high in iron and aluminum and has a clay content of up to 15 percent. Halloysite, the type of clay found in the subsoil eaten by mountain gorillas in Rwanda, is similar to kaolinite, the principal ingredient in Kaopectate, the pharmaceutical commonly used to soothe human gastric ailments. Kaolinite is one of the clays that help reduce the symptoms of diarrhea by absorbing fluids in the intestine. It also adsorbs bacteria and their toxins and could potentially adsorb the higher levels of toxic secondary compounds encountered during the dry season. Other benefits are likely to be gained from this volcanic powder. In the dry season, gorillas have to go farther up the mountain to feed, up to 3,000 meters — heights at which they could conceivably feel the effects of altitude anemia. Researchers infer that gorillas may therefore need the extra iron in the powdered rock as well. Furthermore, the high aluminum content of the powder is likely to provide an antacid effect. The ancient weathered substrates that make up this volcanic rock are completely free of toxic impurities, making it a safe medicine for the gorillas. Soil surveys of the area suggest that the gorillas have found the only available source of their medicine in the Virunga Mountains.[30]

Knowing that animals eat clay, and that clay potentially has these medicinal properties, is not sufficient evidence that the animals are self-medicating — or indeed gaining any incidental health benefit from eating it. We need to know more about the health and diet of the animals that eat clay. Some fascinating evidence has emerged.

Wild chimpanzees at Mahale in Tanzania do not mine and grind rock in the same way as gorillas, but they do take regular mouthfuls of termite-mound soil and scrape subsoils from exposed cliff faces or riverbanks. A chimpanzee passing a termite mound will put out a hand, take a walnut-sized piece of earth, and pop it into its mouth as it goes on its way. When scientists spent 123 hours looking specifically at the health of five chimpanzees eating termite-mound soil, they found that all five were unwell, suffering obvious diarrhea and other signs of gastrointestinal upset. The termite-mound soil at Mahale is low in calcium and sodium, so it cannot be a source of salt or lime for the apes. It is, however, high in clay (up to 30 percent) — more specifically, in the same sort of clay used by mountain gorillas and sold by human chemists to treat gastrointestinal upsets in the West. The termite-mound soils appear to be a source of soothing medicine, used not only by chimpanzees but by many other species (giraffes, elephants, monkeys, rhinoceroses). Many human populations, such as Australian Aborigines, eat termite-mound soil when they have diarrhea.[31]

In the rain forests of the Central African Republic, forest elephants and other mammals have created large treeless licks, ranging in size from 2,000 to 55,000 square meters, many containing holes and caves. These clearings are so well hidden by the dense forest that field workers have to use local BaAka pygmies to find them. All of the licks are on outcrops of ancient subsoils, and most are high in minerals

An African elephant digs down to ancient subsoil for the clay she needs.
*Martin Gruber*

such as sodium, magnesium, potassium, calcium, and manganese. But almost a third of the licks have lower levels of these minerals than surrounding soils. The one feature *all* the sites have in common is a clay content of more than 35 percent.

Evidence that forest elephants are using the clay to self-medicate against gastrointestinal upset is circumstantial but fascinating. These elephants feed primarily on leaves all year long, except for the month of September, when ripening fruit is so abundant that they change to eating mainly fruits. Leaves (as opposed to ripe fruits) generally contain many secondary compounds designed to deter herbivores from feeding on them. A shift from eating leaves to eating fruits would therefore dramatically reduce the consumption of toxic secondary compounds — a natural experiment to see if toxin consumption equates with clay consumption. The only month in which elephants reduce their visits to the clay licks is during that fruit-eating month, September![32]

In Africa, then, mountain gorillas, chimpanzees, monkeys, and forest elephants appear to be eating clay to deal with toxins (or their

Large areas of forest are kept clear of vegetation by the elephants' clay mining. *Gregory Klaus*

effects) in their tropical forest diet. In the forests of South America too, clay consumption is particularly common in parrots, macaws, monkeys, tapirs, peccaries, deer, guans, currassows, and chachalacas. After studying geophagy in the Amazon Forest of Peru for many years, Charles Munn has concluded that nearly all vertebrates that feed on fruits, seeds, and leaves also eat clay. On an average day, he has observed up to nine hundred parrots from twenty-one species, and one hundred large macaws, gather to feed on the eroding riverbanks, biting off and swallowing thumb-sized chunks of orange clay.[33]

In 1999 the hypothesis that macaws eat clay in order to deactivate plant toxins was tested experimentally by James Gilardi and a team of scientists at Davis, California. First, they established that seeds eaten by macaws contain toxic plant alkaloids. Then they fed one group of macaws a mixture of a harmless plant alkaloid (quinidine) plus clay. A second group of macaws were fed just the quinidine, without any clay. Several hours later, the macaws that had eaten the quinidine *with* clay had 60 percent less alkaloid in their blood than

Blue and yellow
macaws in Peru
feed on clay that
binds dietary toxins.
*Jeff Kidston and
Dawn Anderson*

the control group, demonstrating that clay can indeed prevent the
movement of plant alkaloids into the blood.

What surprised the scientists was that the clay remained in the
macaws' guts for more than twelve hours, meaning that a single bout
of geophagy could protect the birds for quite some time. They suspect that clay not only prevents plant toxins from getting into the
blood, but also lines the gut and protects it from the caustic chemical

erosion of seed toxins. As macaws have no diarrheic response to toxins, the consumption of clay may be an essential part of their diet, allowing them to successfully utilize foods that other animals are unable to tolerate.[34]

It is evident that clay is sought by many animals suffering from gastrointestinal malaise. Their discomfort is often caused by plant toxins but can also derive from internal pathogens. In fact, eating clay is used as an *indicator* of gastrointestinal upset in rats. These rodents are unable to vomit, and when they are experimentally poisoned with lithium chloride, they eat clay. This "illness-response behavior" is dose dependent — the more sick the rats feel, the more clay they eat. If they are then given saccharin with the poison, they learn to associate the sweet taste with the feeling of nausea. They will then eat clay even when given saccharin alone. This has led some scientists to conclude that if a rat even *thinks* it has been poisoned, it eats clay. However, the rat does not merely "think" it has been poisoned; because of the conditioning process it *feels* poisoned by the saccharin. Although the difference seems subtle, it may have bearing on our understanding of self-medication. For I propose it is the removal of unpleasant sensations that drives self-medication.[35]

Although geophagy is most common in plant eaters, the fact that meat eaters do occasionally eat dirt is yet another indication that dirt does more than supplement a mineral-deficient vegetarian diet. Tigers occasionally ingest soil deliberately, and George Schaller noted that in India numerous tiger droppings consisted wholly of black micaceous soil, at least during November and December — why these months is not certain. Similarly, the feces of wolves in North America often "appear to have large amounts of earth in them," and although geophagy has not been observed in wolves, certainly their domestic relatives frequently consume dirt, earth, sand, and rocks.[36]

If there is one fact on which scientists researching geophagy agree, it is that the phenomenon has many benefits. The director of the Geophagy Research Unit at York University, William Mahaney, concludes that *all* geophagy is a form of self-medication. And the consumption of soil is so widespread and so inextricably linked to wild health that Timothy Johns suggests that geophagy could be the earliest form of medicine. Although some soils can be a source of nutrients (minerals and/or trace elements), the primary benefit of clay consumption is in countering dietary toxins. In essence, eating earth allows animals to deal with the effects of unavoidable toxins.

## DIRT DISCRIMINATION

Although clay may sound like an excellent medicine, it is not wise to rush out and start eating soil from the garden. Ordinary topsoil may be contaminated with parasite eggs, harmful bacteria, heavy metals, and other pollutants, so it could do more harm than good. When children persistently eat handfuls of soil, the behavior is medically described as pica — an aberration in normal feeding behavior that can lead to anemia, parasitic infection, and even death. Wild animals carefully select the soils they eat, concentrating on ancient, well-leached subsoils that contain fewer and less harmful microorganisms and toxins than topsoils. Even the high-rise soils of termite mounds, widely eaten, consist of subsoils brought to the surface by termites during the construction of their huge homes.

Some human populations retain the knowledge of how to select soils for eating. In West Africa, people eat the *interior* of termite mounds. They also take clay-enriched alluvial soil, from at least 30 to 90 centimeters below the surface, and bake it before eating to be sure it is safe. In Ghana, clay known as eko is extracted from pre-Cambrian sediments exposed in shale quarries, sundried, and distributed to local markets. Kaolin is the major component of eko, making it a highly suitable treatment for gastrointestinal problems. In the United Kingdom, kaolin for the medical treatment of diarrhea is mined in Cornwall.

Although we do not yet know exactly how animals select the right soils, preferred soils are usually yellow or reddish rather than the brown or black of humus-rich topsoil, which is often full of bacteria. Hybrid macaques in Hong Kong look for the finest clays, rejecting soils with larger particles.[37] Parrots and macaws in Amazonia similarly select soils of extremely fine texture from a very narrow band of the eroded riverbed. One characteristic of preferred soils is that they contain at least 10 percent clay, so the texture of the soil could be a valuable cue. Taste does *not* seem to be reliable, as people who eat soils vary in their preferences for salty, sour, or sweet-tasting earth.

## CHARCOAL AS MEDICINE

Charcoal, like clay, can adsorb toxins. Created by heating wood in a restricted air supply, charcoal is capable of adsorbing up to two hundred times its own weight. In hospital emergency rooms it is used as a universal antidote to treat drug overdoses and unidentified poisonings. Doctors also prescribe charcoal for gastrointestinal bloating or wind. Activated charcoal in water purifiers is used to bind and deactivate herbicides and pesticides.

On occasion, wild animals will come across charcoal after forest fires or lightning strikes. Juliette de Baïracli Levy writes that "wild animals know the benefits of eating charcoal and when they scent wood-smoke will come in herds to eat the resultant charcoal."[38] She has seen deer, ponies, and even bees flock to the charred remains of forest fires in the New Forest, England. Other studies by ethologists confirm her observations. During a seven-year study of New Forest ponies, one scientist noted that they will stand around burned gorse bushes after heath fires and browse on the charred branches.[39] Camels are known to eat charcoal in Kenya, and my own domestic horses and dogs eat charcoal from recent bonfires, if given the opportunity. Charcoal has minimal nutritional value and research suggests that animals consume it for its medicinal, toxin-binding properties.

On the African island of Zanzibar, red colobus monkeys in the Jozani Forest Reserve actively forage for charcoal from charred stumps, logs, and branches, and even steal charcoal from the kilns of the local people. Primatologist Thomas Struhsaker has noticed that this particular group of monkeys passes the habit from mother to offspring by imitation. The diet of the charcoal-eating monkeys consists of exotic trees such as Indian almond (*Terminalia catappa*) and mango (*Mangifera indica*), high in phenols and other toxic secondary compounds that could interfere with the monkeys' digestion. Experiments show that the charcoal they eat adsorbs these particular phenols. In particular, the kiln charcoal (which the monkeys prefer) is much more efficient than charred stumps or branches. These monkeys easily consume 2 grams of charcoal per kilogram of body weight — the amount recommended as an effective veterinary dose — an indication that the monkeys may have learned to use charcoal

A red colobus
monkey in Zanzibar
eats charcoal that
combats dietary
toxins.
*Thomas Struhsaker*

to counteract the toxicity of their exotic garden diet. The benefits of
this charcoal-eating habit are becoming all too obvious, as this population of colobus monkeys is reproducing at a higher rate than those
elsewhere.[40]

Although charcoal eating is not as common as clay eating in human societies, it does have a long (and medicinal) history, and charcoal is frequently added to food. Native American Indians pulverized
and mixed charcoal with water, and drank it to alleviate digestive disturbances. Evidence from fossilized feces suggests that prehistoric
Neanderthal hominids also ate charcoal.

Animals are known to eat ash as well as charcoal. Ash eating has
also been seen in wild elephants and domestic livestock, perhaps

serving as an antacid. Chimpanzees in Gombe have on occasion raided fishermen's huts and fed on the remains of their fires.

## COUNTERACTING MAN-MADE POISONS

Since World War II, wild animals have been exposed to greater and greater concentrations of artificial toxins. Most of these pollutants are new chemicals with which animals have no evolutionary history. Not having been exposed to these chemicals over generations, the animals have not been under any evolutionary pressure to develop mechanisms for detecting them. In South Africa, for example, birds (especially birds of prey) have been poisoned by artificial pesticides that the birds consume readily. And marine mammals eagerly consume fish polluted with DDT, dioxins, and PCBs.

Fortunately, certain species with particularly acute chemical detection systems are better able to identify pollutants. Atlantic salmon, for example, will not swim up rivers polluted with copper or zinc mining wastes. Salmon may be exceptional because they have an ability to detect minute differences in water chemistry, which helps them locate their home rivers after a long migration. However, other less sensitive fish species will also move out of areas polluted with aquatic herbicide. When cod and sharks come across unnatural chemicals, such as detergents and dispersants used in crude oil spills, they show their classic escape reactions. So there is hope that some wildlife will be able to avoid at least a portion of the pollution with which we have bedeviled them.[41]

There is interesting evidence of self-medication against the effects of pollution. When rats are fed the lethal pesticide paraquat, they eat montmorillonite clay (fuller's earth), which adsorbs the poison. Poisoned rats that are denied access to clay do not survive. To be effective, though, the clay has to be eaten regularly for many weeks *after* ingestion of the pesticide, to remove all the paraquat from the body.[42] When behavioral responses to toxins are better understood, we may be able to use the frequency of animal self-medication as an indicator of pollution in an ecosystem.

*

We can conclude that familiarity with potential foods is key to avoidance of poisoning. Wildlife managers therefore need to be careful about introducing, relocating, and moving wild animals to unfamiliar areas. Murray Fowler warns that an animal's lack of knowledge about local plants could put it at a severe disadvantage. This is equally important when releasing hand-reared wild animals back into the wild. Early exposure to local poisonous plants is extremely important for young animals that will one day have to discriminate safe from unsafe food on their own. We cannot assume this ability is innate. When avoidance is not possible, some species dilute, combine, or degrade plant toxins before eating them. Others enhance the body's detoxication processes by eating clay, earth, or charcoal to lessen malaise by reducing or preventing the uptake of toxins into the blood. Some appear to speed up expulsion of toxins by eating emetics or laxatives that enhance vomiting and diarrhea.

We can now see why domestic animals fare less well than wild types when it comes to poisons. Livestock particularly are exposed to plants with which they have no evolutionary history. They are often denied an opportunity to familiarize themselves with their environment at a young age, whether through trial and error or by watching their natural mother and peer group. Their pastures often lack the diversity of natural vegetation. This restricts dietary choice; and monocultures are also more prone to diseases caused by bacterial and fungal toxins. Enclosures prevent livestock animals from moving away from plants they might otherwise avoid.

Domesticated animals may still retain some of their natural strategies for dealing with poisoning, as we shall see in Chapter 15. And if these natural tendencies can be encouraged or facilitated rather than hindered, livestock may be better able to help themselves avoid poisoning. For example, it has long been known that cattle benefit from eating clay, and that free-ranging cattle will help themselves to clay, yet in intensive farming they are rarely provided free access to suitable clays. Despite widespread appreciation of the importance of geophagy to many zoo species, captive wild animals are rarely offered similar soils to ingest — despite the high incidence of chronic gastrointestinal problems.

As for humans, we can do our best not to hinder our own efficient vomiting and diarrhea responses to poisoning, and we can learn to make better use of safe, clean charcoal drinks, clay pills, and mineral waters. We live in a world full of potential toxins, many of which are unidentifiable beforehand, so it would be wise to provide clay pills in first-aid kits around the world. What wild animals are showing us is that it is time to overcome our cultural distaste for dirt eating and acknowledge it for what it is: a natural self-medication strategy as old as the hills themselves.

# 6

>|  >|  >|  >|

## MICROSCOPIC FOES

Health consists of having the same diseases as your neighbours.
— Quentin Crisp

BEFORE THE ADVENT of microscopes, people explained the spread of disease by evoking evil mists or winds. Pungent garlands of herbs were hung on doorways and windows to protect against the noxious vapors. Aromatic and antiseptic rue was planted in Arabian gardens to ward off the evil eye, and Native Americans, well aware of contagion, isolated their sick and burned antiseptic incense to ward off bad spirits.[1]

Within a few years of Robert Hooke's initial discovery of the minutiae of everyday objects using early microscopes in the mid-1600s, Anton van Leeuwenhoek became the first scientist to see discrete microscopic organisms. Later, scientists found them in diseased animals and imagined that these tiny objects spontaneously emerged as part of the disease process. It was only in the 1880s that Louis Pasteur and Robert Koch showed how these "germs" caused infectious diseases as they were passed from one organism to another. Thus evil spirits were replaced by disease-causing microorganisms (pathogens). Armed with germ theory, doctors began to use antiseptic techniques in treating patients, and within fifty years powerful antibiotics were discovered that were capable of killing many of these pathogens.[2] With the mechanism of disease transmission understood, and an arsenal of new antibiotics and vaccines, by the

1950s a future free of infectious diseases looked not only possible but likely.

Unfortunately, though, the popular conception of microorganisms as the cause of disease is all too simplistic. A healthy human body contains about 100 trillion microorganisms, many of which are harmless, or even beneficial, in some circumstances but pathogenic in others. Pathogenicity is dependent on a range of factors, not least the conditions inside us. Genetic variation accounts for some of this variation (males are more susceptible to disease than females), but so do personal history and current physiological and psychological status. Diet, hormonal state, age, and lifestyle all contribute to differences in our susceptibility to disease. The neurophysiologist Robert Sapolsky explains in *Why Zebras Don't Get Ulcers* that even something as seemingly innocuous as the number of friends we have can directly affect our ability to fight disease. Social isolation can be as deadly as smoking, obesity, hypertension, or lack of exercise. And regardless of gender, age, or race, poverty is the most powerful determinant of human disease.[3]

Disease is therefore dependent on individual circumstances, and not simply on *invasion* by a specific pathogen. As we saw in Chapter 1, animal health is directly related to environmental conditions. When Cynthia Moss began her studies of wild savannah elephants in Amboseli National Park, she noted that only a few individuals, already in poor health, succumbed to an outbreak of deadly anthrax. The rest of the herd escaped unharmed. Australian marsupials only succumb to disease when they are enduring drought or loss of habitat, and carnivorous marsupials may have no recorded infectious diseases at all in the wild.[4]

Although we have had effective antibiotics for the last fifty years, already species of *Streptococcus, Staphylococcus, Pseudomonas, Enterococcus,* and *Mycobacterium* — so-called superbugs — have developed resistance to all currently available antibiotic therapies. Few new antibiotics are under development. Vaccines too have been useful for preventing infectious diseases such as tetanus and smallpox, but have to be specifically targeted to each pathogen. The number of vaccines is tiny compared to the vast array of pathogens. Furthermore, infectious agents are emerging against which we have no effective defenses.

It is urgent that we find more sustainable ways of fighting infectious disease. One way to improve our chances against the ubiquitous onslaught of microscopic foes is to study the way wild animals deal with them. Although an animal's ability to resist disease is related to the competence of its immune system, we need to know what aspects of their behavior contribute. What do wild animals do to help themselves deal with these invisible health hazards?

## HYGIENIC BEHAVIOR

Even without understanding the nature of disease transmission, animals have evolved many ways of avoiding infection. The honey bee (*Apis mellifera*) is famed for its hygienic behavior, which is clearly under identifiable genetic control. The role of the worker honey bee is to feed the growing brood and keep the nursery cells clean. Workers with a "hygiene gene" can detect diseased brood in the larval or pupal stages, remove the cap from the pertinent cell, and carry the brood out of the nest, thereby reducing the spread of disease. Strains of honey bee that do not have the hygiene gene do not remove infected brood, and infections spread quickly through the hive.

Different strains of honey bee have slightly different abilities in this regard. In the Philippines, nurse honey bees (*Apis cerana*) keep European foulbrood or Thai sacbrood virus infections under control by capping the cells of brood with antiseptic wax. Some strains can even tackle the deadly *Varroa jacobsoni* infection that has in the last century seriously depleted honey production around the world.[5] In Nepal, a strain of bee has been found that protects the whole hive by sealing the caps on cocoons infected with only a *single* varroa mite. Australian beekeepers are exploiting the honey bee's natural predisposition for hygiene by selectively breeding particularly hygienic bees. Enabling these insects to help *themselves* can potentially save the honey industry billions of dollars in lost revenue.

Personal hygiene is a big issue for mammals, too. Jane Goodall writes that "the Gombe chimpanzees, in fact, seem to have an almost instinctive horror of being soiled with excrement." They furiously remove any feces with handfuls of leaves and also wipe themselves clean if soiled with urine, blood, or mud — all potential sources of pathogens. A male will clean his penis after copulation, and a chimpanzee's aversion to feces can even preclude sex. One sick female, Gigi, approached and presented her rump to Hugo, inviting sexual contact. Initially interested, he then noticed and stared at her diarrhea-smeared bottom, changed his mind, and moved away without mating with her. Gigi persisted with other males, until a less fussy (or more desperate) male solved his dilemma by carefully wiping her bottom before proceeding to mate. Natural selection favors animals that successfully avoid pathogen hot spots.[6]

One wonderful advantage of growing up in a stable ecosystem is that youngsters are exposed early to local pathogens and acquire immunity to them. Trouble ensues when strangers bring in foreign pathogens from far-off places. The measles virus, to which most of us in the West have some natural immunity, is as destructive as biological warfare to communities with no previous exposure to it. Currently, the last of the Jarawa tribe of the Andaman Islands are being eradicated by a measles epidemic caused by strangers building a new road through their lands.[7]

The presence of strangers has always been a threat to animal populations, and the natural fear of strangers shown by animals (including ourselves) may be based on the experience that disease and death often follow. Certainly, chimpanzees show an innate distrust of contact with strangers. Goodall once observed a female chimpanzee approach a male of another group and reach out her hand to touch him. He immediately moved away, picked some leaves, and wiped where she had touched his skin. Another time, a youngster investigating an unfamiliar human visitor stamped on the visitor's head with her foot. She then sniffed her foot, picked some leaves and cleaned it vigorously. Fear of the unfamiliar is helpful in avoiding contact with foreign pathogens.

Cannibalism is not as common as it might be, considering the obvious nutritional and competitive advantages of eating other members of one's own species (conspecifics). Its rarity can be explained by a strong selection for those that do not eat close relatives — but another possible factor is disease avoidance. Conspecifics share a similar genetic susceptibility to certain pathogens, thus are likely to harbor pathogens specifically adapted to the cannibal's own physiology.

Furthermore, pathogens can mutate and adapt to the conditions in which they find themselves, meaning that pathogens incubating in a conspecific are even *more* dangerous than those incubating in the body of a different species.

This assumption was tested in 1998 by David Pfennig. Tiger salamander tadpoles readily feed on tadpoles — usually those of other species. Pfennig fed healthy tadpoles a selection of four diets: infected or healthy tadpoles of their own species, and infected or healthy tadpoles of a different species. Only tadpoles that ate diseased conspecifics failed to thrive, even though the pathogens were the same in all diets. Eating one's *own* dead seems to be more dangerous than eating the dead of other species![8]

In most human societies, cannibalism is considered revolting — a reaction that is highly adaptive. In those cases in which ritual cannibalism has been documented, it has been found to adversely affect the cannibals' health. In Papua New Guinea, the Eastern Highlands Fore people suffered from a mysterious disease called kuru, which attacked the nervous system and sent individuals into fatal fits. In the 1950s the cause was discovered to be a prion infection (a protein mutation that spreads in brain and spinal-cord tissue) transmitted through eating human brains. Women had ritually eaten the brains of loved ones in an honorary gesture as they prepared them for burial.[9]

Bovine spongiform encephalopathy (BSE), or mad cow disease, is caused by feeding not just meat to a vegetarian animal but meat *of its own species* — ground-up bits of cattle put into the cattle feed. Again the causal agent is a prion, which has been passed along the food chain to humans in the form of a new variant of Creutzfeldt-Jakob disease (nvCJD). Both BSE and nvCJD are fatal. If the lessons of disease avoidance demonstrated by wild animals (and by our own behavior) had been observed, mad cow disease and its nvCJD offshoot could have been avoided.

## REACTIONS TO INFECTION

When avoidance fails and pathogens gain a hold, there is still much an animal can do to help itself. One of the more subtle approaches is the behavioral manipulation of body temperature. Microorganisms can be killed by high temperatures, but so can many vital enzymes in the body. So to rid the body of infection, animals need to hold body temperature at an optimum level — not too high for their own delicate enzymes, but high enough to harm pathogens. So-called cold-blooded animals that actively moderate their own body temperature (such as tortoises, toads, and crabs) need to take direct action to bring on a germ-killing fever. Sick lizards seek out warmer places to sit, thereby raising their body temperatures by the two degrees needed to kill pathogens. If they are unable to find a warmer place, they are more likely to die. Fish also move to warmer waters when experimentally injected with bacteria.[10]

Even insects actively increase their body temperature to deal with an infection. Ants and flies infected with fungus crawl up vegetation and bask in the sun, and crickets and grasshoppers infected with bacteria seek warmer places that increase their chances of survival. Eusocial insects take group action: when infected with a fungus known as chalk brood, infective at low temperatures, honey bees vibrate their large wing muscles, raising the temperature of the hive just high enough to prevent brood larvae from being infected, apparently detecting the pathogen before symptoms appear.[11]

Mammals and birds, with their internally modulated body temperatures, automatically develop a pathogen-killing fever when an infection takes hold. A fever is a defense against infection, not a symptom to be suppressed. In an experimental trial, children with chickenpox who were given fever-reducing medication took a whole day longer to recover than those given a placebo sugar pill.[12] Warm-blooded animals therefore do not necessarily have to seek out warmer places to recuperate. Nonetheless, many do sun themselves when sick. When the Welsh Hedgehog Hospital gets a phone call from someone saying a hedgehog is sunbathing on their patio, they know the hedgehog is ill. In addition to raising body temperature, it could be that sunlight acts as a direct antimicrobial agent. Ultraviolet light alters the DNA of many microorganisms, rendering them unable to reproduce, and it is lethal for most bacteria, fungal spores, viruses, protozoa, nematode eggs, and algae. This is why ultraviolet light is used to treat water for human consumption.

Often the first indication of sickness is that an animal goes off its food — an observation that has led many traditional herbalists to conclude that fasting is a natural and beneficial response to sickness. Juliette de Baïracli Levy writes, "A sick animal retires to a secluded place and fasts until its body is restored to normal."¹¹ This lack of appetite can be an effective way of speeding recovery. Bacteria need iron, and the body makes many adjustments to reduce the availability of iron during an infection. If an animal keeps eating, any iron in its food also feeds the pathogens, so force-feeding a sick animal or patient can be counterproductive. Fever reduces appetite, so if a fever is artificially suppressed with drugs and appetite returns, not only do we lower temperature but the increased iron intake can keep the infection going longer. The traditional dictum "Starve a fever" is medically sound advice.

Vomiting, as we saw in the last chapter, is an effective way of expelling toxins and pathogens. Pet owners may have seen their dog or cat chew grass and soon afterward retch up the contents of its stomach. In this case, the grass (often couch grass) is acting as an emetic. Diarrhea also expels pathogens and poisons, yet often we hinder this vital function by stopping the first sign of "holiday diarrhea" with strong medication. In tests, people given an antidiarrheal drug, Lomotil, took longer to rid themselves of infection than people who did not get the drug.[14] In other words, sometimes *not* taking medication is the best way to stay well. The animal responses to infection — fasting, resting, staying warm, and allowing vomiting and diarrhea to do their jobs — are tried and tested by natural selection. We hinder them at our peril.

## DIETARY PREVENTION

Most of the microorganisms in the body reside in the gut, and health is closely related to the condition and relative numbers of gut microflora. About a third of the dry weight of human feces consists of the dead bodies of the four hundred to five hundred species of bacteria living in the gut. This cocktail of bacteria is carefully balanced by gut conditions, and these in turn are distinctly affected by diet. Some of the bacteria make essential vitamins, others support the immune system, and some aid food digestion. These so-called facultative bacteria can be so important to our health that we become ill if they are not present. The bacterium *Bifidobacterium bifidum,* for example, not only keeps pathogenic bacteria in order by competitive exclusion (suppressing the growth and division of other bacteria) but also helps prevent colon cancer. Diet can directly influence competitive exclusion. Laboratory research on rats shows that calcium phosphate supplements in the diet can feed the beneficial *Lactobacilli* that fight off pathogenic *Salmonella* bacteria and thereby protect rats from gastrointestinal infections.[15]

Within the first week of life, an infant mammal gets its complement of *Bifidobacteria* from its mother's milk. Breast milk is more than mere food: it has been known for some time to contain antibodies from the mother's immune system that help the infant combat diseases the mother has already encountered. More recently, it has been found that breast milk also contains medicinal compounds that help the infant fight other infections.[16]

Breast milk is therefore the first medicine a newborn mammal consumes. Once weaned, young animals continue to obtain boosters of their mother's resident microflora from any food she may chew for them. Many youngsters also top up their gut microflora by eating their mother's feces. As soon as young foals are born to feral mares in the New Forest, they start to nibble at their mothers' dung. This habit gradually subsides and stops after three months.[17]

For many herbivores, bacteria are particularly useful for breaking down tough plant cells, allowing access to nutrients. Eating feces (coprophagy) is one way animals such as gorillas, elephants, rabbits, and hares add to their supply of essential bacteria through adult life. It is therefore an important aspect of their health care. Overzealous rabbit owners could inadvertently cause disease in their pets by cleaning out the droppings too quickly. By reingesting their own soft pellets (and discarding the hard, dry ones), rabbits extract further nutrients from their food and obtain essential vitamins made in the gut by microorganisms.

Modern broad-spectrum antibiotics kill bacteria, good and bad. Even more specific antibiotics disrupt the balance of resident microorganisms, perhaps allowing a pathogen that was previously kept in check to take over. When whiteflies are given antibiotics, they soon show signs of reduced growth, they stop reproducing, and some die.[18] Antibiotics therefore need to be administered with great restraint. After a course of antibiotics, when gut microflora are disrupted, we do not (luckily) have to resort to coprophagy. We can help rebalance the cocktail of bacteria by eating foods such as live yogurt that contain beneficial *Bifidobacteria* and *Lactobacilli*.

We have seen that plants contain many natural antimicrobials. Do these not harm the gut microflora of wild animals in the same way as antibiotics? In excess they do (as we saw in Chapter 5), but as part of the normal diet they may play an essential role. Herbivores have diets particularly rich in natural antimicrobials. Over a third of the leaves, fruit, bark, fruit, seeds, and pith that make up a mountain gorilla's diet contain antimicrobials. John Berry, a phytochemist, has discovered that these do not disturb beneficial bacteria but do keep potentially harmful bacteria in check. Up to 90 percent of the diet of lowland gorillas in western Africa is made up of the fruits of *Aframomum* (a plant of the ginger family), which contain antimicrobial compounds that help gorillas maintain a healthy balance of beneficial and harmful bacteria in the gut.

Bacteria are classified into two distinct types, gram negative and gram positive (depending on how they are stained by a dye named after the Danish scientist Cristof Gram). The two types have slightly different characteristics. Many pathogenic bacteria such as *Salmonella* and *Shigella* are gram negative, and it is these that are most susceptible to the antimicrobials in gorillas' food. The gorillas' natural diet therefore keeps pathogens under control while allowing their gut flora to remain healthy. Captive gorillas might benefit from inclusion of these natural antimicrobials in their diet, as they commonly suffer serious bacterial gut infections such as shigellosis, which can lead over time to arthritis.

In addition to consuming a diet rich in antimicrobials (*Aframomum* is just one such example) that are prophylactic against infections, sick gorillas appear to take some curative measures. Berry reports that the mountain gorillas of Mgahinga National Park, Uganda, eat the bark of *Dombeya quinqueseta* when suffering from diarrhea. In the laboratory, this bark inhibits the growth of *Salmonella* — pathogenic bacteria that commonly cause diarrhea. Local people also claim that sick mountain gorillas climb into alpine regions to eat the leaves and roots of lobelia plants, which are used by herbalists to treat bronchitis, colds, and coughs.[19]

## BOOSTING PROTECTION

Food does more than affect the ratio of beneficial to harmful bacteria in the body. It does not simply supply the immune system with the nutrients it needs to function (mammals deficient in zinc, calcium phosphate, and vitamins are more prone to infection); it has subtle immune effects that can pass unnoticed in animals not obviously malnourished. There is strong variation, for example, in the susceptibility of Colorado potato beetles to fungal infection when the beetles are reared on different host plants. Similarly, palm oil can protect mice against *Listeria* deaths.[20] Exploring the link between diet and disease resistance might help us to improve our own health care.

After the Chernobyl nuclear catastrophe, Russian phytochemists intensified their search for plant compounds that might protect people against opportunistic infections. At Moscow's Institute of Medicinal Plants scientists screened numerous plants and found that glycoalkaloids such as solanine and chaconine in *Solanum* plants could boost the immune systems of mice. These same glycoalkaloids at high concentrations are toxic, causing liver dysfunction and even death, so their protective role at lower doses was a surprise.

Two of these chemists, Michael Gubarev and Elena Enioutina, later discovered something even more unexpected. They found that it took only a *single* low dose of solanine to effectively protect mice from infection with *Salmonella* bacteria, and the protection from that one shot lasted for as much as two weeks. In comparison, mice not given solanine died within four days of *Salmonella* infection. Solanine has absolutely no effect on bacteria in the test tube; so it is *not* an antibacterial agent. It only works inside the body, by helping the natural immune response. The mechanism is not fully understood, but the blood of solanine-treated mice contains an unidentified component that helps destroy bacteria.[21]

In the past, it has not always been easy to explain why so many animals — from maned wolves to baboons, rhinos to insects — regularly consume potentially toxic *Solanum* plants. Some insects (as we shall see) fight pathogens using this toxicity, but we should not ignore the potential protection afforded by immune enhancement. Stimulation of the immune system by small amounts of toxin is an example of the hormetic effects mentioned in Chapter 5. If one toxin can do this, probably many others can do the same. Hormetic effects are likely to be an important factor in the health of wild animals.

We should also consider the hygiene hypothesis put forward to explain the phenomenal rise in childhood allergies in the West over recent decades. The hypothesis is that our immune system has evolved to be ready to fight a range of infections in early childhood and that, when an overly hygienic lifestyle reduces exposure to childhood infections, the immune system goes looking for trouble, fighting off innocuous substances such as house dust, food, or even our own tissues.

This theory may explain the recent rise in autoimmune diseases in the West. Scientists have found that inoculating people with small doses of soil bacteria (mycobacteria, similar to those that cause tuberculosis and leprosy) can stimulate the immune system and thereby reduce tumor growth. Diverting the immune system into action against a real pathogen in this way has been found to alleviate symptoms of multiple sclerosis, diabetes, and allergic asthma.[22] In other words, we may *need* to be exposed to infections in early life to keep our immune system working efficiently. Young wild animals are of course exposed to a vast range of microscopic foes that keep their immune systems busy and may thereby prevent autoimmune problems in later life.

To carry our discussion one step further, are there any examples of animals actively seeking antimicrobials to prevent or treat infection? Nature's pharmacy is certainly full of antibacterial, antifungal, and antiviral compounds. While observing their effects on animals is difficult, we do know that animals gain benefit from natural antimicrobials.

The honey bee is a master chemist when it comes to fighting infections. Bees collect resins produced by trees to protect their vulnerable buds from weather and pathogens. From these resins, bees make

A spotted cucumber beetle obtains bitter cucurbitacins, which protect its eggs and larvae from infection by soil fungi. *Douglas Tallamy*

propolis (a Greek word meaning "defender of the city"), with which they coat the entire inner surface of the hive, producing one of the most sterile environments in nature. Propolis provides structural support as well. Because the precise type of resin used by the bees varies, the exact chemistry of propolis varies too; but commonly it contains hundreds of different flavonoids, phenolics, and aromatic compounds. If an invader — say, a rat — enters the hive and dies, the bees embalm the corpse with sticky propolis that seals off and protects the hive from harmful microorganisms.[23] The honey they make as a storage food for the winter is also packed with potent antimicrobials that help prevent spoilage. It is known that bees seek out aromatic antiseptic plant compounds, and traditional herbalists use both propolis and honey to stimulate the immune system and treat wounds.

Other insects also make use of natural antimicrobials. Gall wasps lay their eggs near leaves with the highest possible tannin concentrations, gaining protection from infection for their emerging larvae.[24] Similarly, when egg laying, spotted cucumber beetles (which as larvae are known as southern corn worms) leave their host corn plants and seek out squash, gourd, or cucumber plants to harvest their bitter secondary compounds, cucurbitacins, which are passed to the beetle eggs. Experiments show that both the eggs and the larvae are thereby protected from soil fungi and predators. It is thought that the beetles use these two plants because for thousands of years Central American Indians planted corn and squash plants together. At least one fungus-growing ant (*Cyphomyrax minutes*) farms a fungus that secretes antibiotics preventing infection of its shared food.[25] In the sea, the Spanish dancer nudibranch (a sort of sea slug) obtains a toxin (kabiramide C) from its diet of sponges and stores it in its egg masses. The abandoned strings of eggs are thereby chemically protected from predators and fungal infection.[26]

Our own attempts at biological pest control have revealed a possibly widespread use of protective plant antimicrobials. Attempts to fight the tobacco hornworm, an insect that feeds specifically on *Solanum* plants (tobacco and tomato, for instance), by spraying with a lethal bacterium (*Bacillus thuringiensis*) are undermined by the hornworm's consumption of large amounts of the antibacterial alkaloid nicotine. This suggests that attempts to *deter* herbivorous insect pests by breeding plants containing greater concentrations of toxic secondary compounds may be fundamentally flawed and even counterproductive, as the compounds can actually protect the insect pests from infections.[27]

According to folklore, the horse chestnut tree was so named because greedy horses fed avidly on the large seeds, curing themselves of chest infections.[28] However, scientific evidence of curative use of antimicrobials is difficult to obtain. One possible example is the consumption of the termite-mound soil by wild chimpanzees (and many other animals). This soil is woven together by filamentous bacteria (actinomycetes) — a group of soil organisms that synthesize about three quarters of all known medicinal antibiotics. Researchers wonder whether sick chimpanzees may therefore be gaining antibiotics with their medicinal clay.[29]

Unaware of microscopic foes, animals are presumably dealing with the symptoms of infection (seeking relief from malaise) rather than targeting pathogens. One common symptom is the copious production of phlegm to protect vulnerable mucous membranes. When mucus blocks their ears, mandrills use ear probes to clean them out.[30] Chimpanzees are even more adept at using tools to relieve their symptoms of infection. In one outbreak of influenza among the chimpanzees of the Mahale Mountains in Tanzania, an adult male, Kalunde, had the flu for two weeks. His nose was so blocked he could breathe only through his mouth. In an attempt to clear his nasal passages, he cleverly made a tool by picking up a large dry leaf, removing the leaf blade with his teeth, and pushing the midrib into one nostril. Within five seconds, he sneezed and released a large amount of mucus onto his upper lip. He continued to stimulate sneezing, using slender twigs, grass stems, or similar tools, throughout his illness. The observers were convinced of his clear intention to induce sneezing and thereby relieve discomfort.[31]

Another distracting symptom of bacterial infection is tooth decay, or caries, caused by bacteria such as *Streptococcus mutans* that feed on sugary food residues. Kenneth Glander, director of primate research at Duke Primate Center, studied the teeth of more than 950 mantled howler monkeys in South America and found a complete absence of cavities and gum disease. These monkeys not only have a low-sugar diet, they also eat a great quantity of cashew pedicels (*Anacardium occidentale*) — which contain the phenolic compounds anacardic acid and cardol that are active against tooth-decay bacteria. This diet of low sugar and high phenols may be helping to protect the monkeys against tooth decay. Phenols such as tannins are common in many plants, and are known to inhibit the growth of *Streptococcus mutans,* a main protagonist in tooth decay.[32]

Chimpanzees, which eat far more sugary fruit than howler monkeys, suffer from both tooth decay and gum disease. To cope with this ailment, they chew on antibacterial barks — which local people also use to keep teeth healthy — and inspect and clean each other's teeth. In captivity, one chimpanzee was even seen to pry out the bad teeth of another by means of a simple wooden lever she had made.[33]

## PATHOGENS IN FOOD

A better knowledge of how animals protect themselves from food poisoning could help us protect ourselves. For example, residents of Lanarkshire, Scotland, experienced an outbreak of a particularly nasty, and lethal, case of food poisoning originating in meat from one supplier. The deadly pathogen turned out to be a form of *Escherichia coli*, which normally lives quite harmlessly in the gut. But this form (called *E. coli* 0157) is different. First discovered in the 1950s and nicknamed the "hamburger disease," it causes bloody diarrhea if you are lucky, and kidney failure leading to death if you are not. It first emerged in intensively reared cattle, and the suspicions are that it has resulted from the excessive use of antibiotics as growth promoters and prophylactic medication. Long-term use of antibiotics disrupts the gut flora and harms microorganisms such as *Bifidobacteria*, which normally keep *E. coli* 0157 in check.[34]

Given the opportunity to choose their diet, animals can apparently rebalance gut microflora and clear themselves of this potentially lethal pathogen. Sheep experimentally dosed with *E. coli* 0157 completely cleared themselves of the pathogen within fifteen days if allowed outside onto a sagebrush-bunchgrass range, yet remained infected if kept indoors on standard grain feed (the role of diet selection is as yet unknown). In 1998 a report in *Science* suggested that to reduce *E. coli* 0157 infection in cattle, hay is better than grain; but grass is better still. Precisely how diet affects *E. coli* in livestock is the focus of current research.[35]

While we humans focus solely on the destruction of pathogens, animals combat infectious disease via a holistic approach that involves avoidance, prevention, and treatment of symptoms. There is no need to imagine that they do this consciously or intentionally; the strategies need only to have been adaptive in the past to have become part of the behavioral repertoire. It is because they have stood the test of time that they are crucial to us now. Our optimism about eliminating infectious diseases with pathogen-targeted antibiotics has proved to be misplaced. Many of the diseases initially controlled by antibiotics have returned with more resistance and greater virulence.

We cannot eradicate all disease. As each new drug is found, resistance to it evolves, and the search for improved pharmaceuticals continues ad infinitum. By adopting the holistic strategies demonstrated by successful survivors in the wild, we can greatly improve our chances of sustaining health. Even these strategies, though, will not always be successful. Individual animals that seem to do all the right things can still die of disease. This is natural. In the arms race between organisms there will always be winners and losers. It is our *expectations* of conquering disease that are unrealistic, not the idea that animals can teach us how to stay well.

✄ ✄ ✄ ✄

# 7

# GAPING WOUNDS
# AND BROKEN BONES

As soon as there is life there is danger.
— Ralph Waldo Emerson, 1860

"THE RECUPERATIVE POWERS of gorillas never cease to amaze me," wrote Dian Fossey as she examined the skulls of two silverback males she had found in the Virunga Mountains of Rwanda. Embedded in the eyebrow bones of each was a piece of tooth from another silverback, and from the bone growth around them she could tell that the gorillas had received these severe wounds in their youth, many years earlier. These are common injuries for gorillas. Of sixty-four skeletons she had collected over the years, 74 percent of silverbacks had healed head wounds and 80 percent had missing or broken canines. In her thirteen years of study she saw many signs of their ability to survive injury, including vast networks of scars and healed gashes that zigzagged their massive heads.[1]

Similar examples of effective healing are common among the chimpanzees of Gombe. An adolescent chimpanzee, Sherry, received a deep gash on her thigh, which stank of putrefaction for a week and was still "visibly bad" after twenty days. Yet within a month the wound appeared completely healed. Another adolescent male was injured by a bush pig during a meat hunt, receiving deep wounds in his lower back. Although he had difficulty walking for five days, he was fully recovered in only two weeks. Plains zebras too have been observed to recover from severe injuries. One mare with a fresh gaping wound, 10 centimeters long and 40 centimeters wide, on her right hip was spotted by ethologists. The wound later became infected and purulent. Three months later, though, the wound had healed completely and the scar could be seen only as a slight displacement in the stripe pattern.[2]

Bones get broken, too — and heal straight and true. A third of Asia's gibbons have at some time fractured a bone. In Mikumi National Park, Tanzania, nearly all yellow baboons more than thirteen years old have old fractures somewhere, and an analysis of dead coyotes and wolves in Canada revealed numerous healed ribs — fractured by kicks they had received from their prey, caribou and reindeer.[3] Even the loss of a hand or foot (usually from a hunter's snare) is not necessarily fatal to a baboon, gorilla, or forest chimpanzee. Most long-term studies of primates include at least one surviving amputee among their subjects.

Somehow these mammals, although they live in pathogen-rich environments, are able to heal cleanly even from injuries inflicted by germ-encrusted teeth or claws. Yet they have no vaccinations against tetanus, no stitching, and no plaster casts, bandages, or slings. Such healing ability seems fantastic compared to our festering ulcers, bed sores, and septicemia.

## STAYING OUT OF TROUBLE

Accidents happen. In the caves of Mount Elgon, Kenya, lie numerous bones of adult and baby elephants that fell to their deaths down deep underground crevasses in their quest for mineral salts. In one year at Gombe, fifty-one chimpanzees were seen falling from trees; two died as a result.[4] Still, animals have evolved many ways of reducing the likelihood of accidents — play being the most enjoyable. It is during play that animals learn about the intricacies of their physical environment, about their own abilities and those of others. Plopping into a fast-flowing river, a young elephant learns just how slippery mud can be. If it is lucky, it is still small enough to be rescued by a single uplifting movement of the mother's trunk. Stretching also helps prevent injury by loosening muscles and eliminating waste products of muscle metabolism. Many yoga movements are named after the favorite stretches of particular animals. The well-known yoga position "downward dog" regularly taken by wolves and big cats stretches all the major hunting muscles along the dorsal and ventral surfaces of the torso and legs.

When big cats scratch at trees, they stretch tendons attached to their retractile claws, keeping them ready for the next kill. Those of us with pet cats know only too well how damaging these scratch marks can be to furniture. The favorite scratching trees of jaguars are clearly identifiable in the jungle from scratch marks reaching 3 meters above the ground. The jaguar hunter Tony de Almeida noticed that in the northern swamps of the Matto Grosso, jaguars scratch exclusively on the "morcegueira" tree (*Andira inermis*), a hardwood with a thick trunk and rough bark.[5] It may be that this is the only species that affords the angle and tactile properties required by the jaguar, or the tree may play a more medicinal role. Its bark contains berberine, a highly bactericidal alkaloid that is highly prized by herbalists. Scratching berberine-rich bark could be an effective strategy to keep those vitally important pads and claw sockets hygienically clean.

## BOYS WILL BE BOYS

In many species, males are injured while competing for females and food. In Tanzania a male yellow baboon can expect to be seriously injured by another male approximately once every six weeks and will take about three weeks to recover from each injury.[6] Luckily, these alarming statistics are not characteristic of primates in general. Baboons have a bit of a reputation for aggression: even leopards and lions (not normally known for their timidity) steer clear of them.

Healing takes time and energy, and an individual waiting for injuries to improve can lose social status, access to mates, food, resources, and territory. Because injuries incur a cost, they have received a great deal of evolutionary attention. On the rugged Hebridean islands of Scotland, red deer stags during rut fight with tenacity over territory and females. However, a stag is only likely to become seriously injured once in its lifetime. There are few mindless thugs in the wild: neither contender would benefit from injury, so both need to be able to weigh the costs and benefits of fighting over a particular resource. Is she or it worth fighting for? How likely is victory? Before risking combat, stags walk up and down, parallel to each other, roaring and bellowing, assessing their opponent's strength and size before making any moves. Usually one of the stags will back down at this stage, disarming further aggression by making a swift exit. However, if the stags are well matched, the situation may escalate into head-to-head combat in which serious injury and even death can result.

## UBIQUITOUS DANGERS

When predators attack prey, both the hunter and the hunted are at risk. A predator will often select the weakest animal and will usually give up the hunt at any threat of injury to itself. Prey species in turn have a wealth of behavioral strategies to avoid being eaten. These often employ substances from nature's pharmacy. Insects, mollusks, amphibians, even birds, harvest toxins from their diet and store them in the outer parts of their bodies to deter predators.

Nature's pharmacy, though, has little effect on one of the most effective predators — humans. During many months of field research in the jungles of southern Mexico I never once saw a jaguar. I found fresh tracks and smelled fresh odor signals. I was followed through the undergrowth one morning by a large male, but I never actually came face to face with one in the wild (in light of my solitary excursions, I must say this was more of a relief than a disappointment). Staying away from humans is probably the most adaptive behavioral strategy that any large mammal enskinned in a popular fashion accessory could have, and generations of hunting have left jaguars wary of man.

Animals cannot so easily avoid other aspects of our lifestyles. Electric cables, barbed-wire fences, and fishing nets are just a few human artifacts that inflict lethal injuries on wild animals who have had little time to adapt to their presence. Not only are road accidents the primary cause of injury to human males, they are also a major threat to wildlife. In Mikumi National Park in 1995, for example, at least three road deaths occurred each day, for an average of 218 road deaths per kilometer of road each year. The animals that were regularly killed included vulnerable and endangered species such as elephants and African hunting dogs. Road deaths accounted for 10 percent of the total yearly losses from the resident yellow baboon troop.[7]

## DEALING WITH WOUNDS

We know surprisingly little about how animals behave when they are wounded. Injury is often first noticed as a stiff gait or lack of movement. A natural physiological response to serious injury is that the area becomes temporarily paralyzed so that the wound or bone break is not worsened. A badly hurt animal often has little choice other than to rest and recuperate.

Another physiological response to injury is pain, but only recently have scientists accepted that animals even feel pain. In 1985 the British Veterinary Association held a seminal conference on recognizing, assessing, and alleviating pain in animals, and in 1987 a similar conference was held in the United States. Before that, it had only been generally acknowledged that animals show the physiological correlates of pain, such as nerve impulses, changes in hormone levels, and reduced movement. It was known that all vertebrates share similar pain sensors, called nociceptors, as well as strong endorphins (internal opioids) that relieve the pain of exertion; but the denial of animal emotions or feelings precluded the concept that animals *feel* pain. This, along with the fear of being accused of anthropomorphism, left many scientists in the paradoxical position of studying animal pain in laboratory experiments, in order to advance human pain management, but at the same time not acknowledging that animals suffered.

Pain is a warning signal. Its function is to prompt the subject to take action to regain health, at which point the pain signal will go away. People born with congenital lack of pain perception often are unaware of extensive injury and die an early death. In other words, pain is an integral part of a health maintenance system. On the other hand, chronic pain can reduce fitness by distracting an animal from essential tasks.

One reason it has been easy to deny that animals feel pain is that they do not always display their discomfort in the same way as humans. A horse in intense pain may reveal its agony only by subtle postural changes: a slight wrinkling of the nose or depression of the eyelid. A wounded fox will continue to run, without yelping, from pursuing hunters. A dog with a painful tooth abscess may remain still and quiet. This lack of demonstration can be interpreted as an indication that animals do not feel pain in the same way we do. However, Bernard Rollin writes in *The Unheeded Cry* that the *perception* of pain is different from the *demonstration* of pain.[8] There is often no benefit to demonstrating pain and many excellent reasons for hiding it. Wounded prey animals conceal their injuries to avoid the attention of predators. To survive, an animal must not only *be* healthy but must *look* healthy to others.

Captive animals often conceal their injuries so effectively that keepers and veterinarians are unaware of them until the animals die of complications. Birds are particularly adept at hiding injuries and often surprise their keepers by dropping dead. Similarly, sheep farmers report that sheep "just die" with little warning. Any sign of weakness in a high-ranking lion, stag, rabbit, or wolf would lead to a takeover bid by a younger, healthier animal, and demotion is usually accompanied by a massive reduction in fitness as access to mates dries up. Injured male olive baboons certainly lose status and females, as do many other primates. If an injury can be concealed while it heals, loss of status may be avoided. But by effectively concealing pain, animals have convinced humans that they either do not feel it or can tolerate it much better than we can. As a result, we may have missed noticing their self-help strategies.

Until the 1980s, there was no universal standard for measuring pain, but one ingeniously simple experiment showed how pain can be measured objectively in nonhumans. Rats with a form of painful arthritis were offered food containing the painkiller suprofen. They ate it readily, preferring it over other food, while rats without arthritis did not. Arthritic rats therefore *demonstrated* a pain that was otherwise undetectable. By the early 1990s, self-selection of analgesic was widely accepted as a way of assessing pain in animals and was used to investigate the efficacy of painkilling drugs. Paradoxically, although animals have proved that they do feel pain, their ability to do so is now used in medical research in which they endure more pain. However, there is some progress: in 1997 European Union law somewhat tardily recognized farm animals to be living creatures capable of feeling pain and suffering.[9]

The welfare of animals in intensive farming is a contentious issue, and any objective measure of their suffering is useful in the debate. A team of veterinary scientists at Bristol University have used chickens' ability to self-medicate as proof that they suffer pain. Broiler chickens have been artificially selected to grow extremely quickly, turning food into meat at the expense of bone growth. Their legs therefore are not strong enough to support their weight, and they frequently suffer broken leg bones; yet they receive no analgesics. Lame birds go off their food and remain still, unwilling to walk — even to the water trough. However, month-old birds can rapidly learn to select feed containing the painkilling analgesic carprofen, and the amount of painkiller the birds eat increases with the severity of lameness. Carprofen tastes slightly peppery (to a human) and can cause gastrointestinal upset. Sound birds tend to avoid the drugged feed, suggesting that they find it unpleasant — an additional indication that the lame birds prefer the distasteful food for its analgesic properties.[10]

Unfortunately this groundbreaking study, showing that lame broiler chickens do suffer chronic pain and that they are able to self-medicate against it, has not helped the chickens. Meat sold for consumption in the United Kingdom must come from animals clear of drugs for twenty-eight days before slaughter, and these birds do not live much longer than that.

If laboratory rats and intensively reared broiler chickens can self-medicate against pain, using drugs they have never come across before, there is no reason to suppose that their wild counterparts cannot do the same with natural analgesic compounds freely available in their environment. Investigating this thesis in the wild is going to be very difficult, not least because of the concealment of pain, but anecdotal observations point to some interesting possibilities (as we shall see in Chapters 8 and 10).

## HEALING RETREAT

Animals that live in social groups cannot always successfully conceal injury from their peers, and they commonly break away to spend time alone. Red deer leave the herd while they are recovering, and injured baboons tend to stay on the periphery of the group, interacting far less with others. Sometimes it is not the animal's decision to leave. Injured male baboons are often persecuted by other males, wounded lethargic reindeer are driven off by other reindeer, and injured badgers are driven away from the sett to hide in some very odd places: barns, outbuildings, lambing pens, even public lavatories.

Wounded wolves also go into hiding. Biologist David Mech tells of nine wolves that were shot by hunters; the solitary survivor ran off and hid under trees. After seventeen days in hiding, he was flushed out and shot again — *but* his original wounds had healed, and the virgin snow around his hiding place showed that he had not moved during all that time.[11]

Injured savannah elephants make their way to swamps with shade, water, and soft vegetation. Cynthia Moss describes the behavior of one matriarch wounded by Masai hunting spears:

> Two of the spears had fallen out, but the first, the one that had gone into her shoulder, was still embedded. It scraped against the bone as she tried to run. She limped her way to the thickest part of the swamp edge and hid among the dense bush. There was no way she could catch up to the others, who were now halfway back to Olodo Are, running at full speed and very frightened. When Teresia reached the thicket she stopped, her sides heaving with the effort. With her trunk she grasped the spear and pulled and twisted it until it came out.

After two days of hiding while recovering her strength, she moved down to the swamp to drink and even feed for a while, but she was not strong enough to try to make it back to the park and her family.[12]

Not all group animals need to isolate themselves when injured. Even serious injury does not necessarily result in a loss of status, especially if the victim has active supporters. One fight among gorillas left a forty-two-year-old silverback with a broken arm. Dian Fossey could see the end of his upper arm bone, surrounded by exposed ligaments and fascia protruding through the skin of his elbow. His son, fourteen years old, had eight deep bite wounds on his arms and head. For several weeks, the two males lay together during the group's lengthy day-resting sessions. The son's wounds healed quickly. Soon the father was left recuperating on his own, on the edge of his group, where he would sleep for most of the day. It took him six months to recover fully, but when he did, he resumed the leadership of his group. There were times when Fossey had doubted his survival, "especially when the wound was draining copious amounts of foul-smelling exudate that attracted scores of insects to his body. Because of the location of the wound on his elbow he was unable to cleanse it orally, which was undoubtedly one reason it had taken so long to heal."[13]

## LICKING OUR WOUNDS

Hiding is by its very nature difficult to observe, so there is a paucity of documentation on the behavior of injured animals. Apparently some species tend their wounds, while others leave them to heal by themselves. Wound tending is commonly reported as "grooming" by observers. The wounds of adult male baboons are carefully groomed by females, and this, along with self-grooming and licking, helps the healing process. Many species, such as primates, canids, felids, and rodents, lick wounds. In most species studied, wound licking is an innate response to injury, not one learned from observation of others.[14] If they can reach them, chimpanzees will lick their own wounds; if not, they will lick their fingers and then dab at the wounds. They also dab leaves on a wound, lick the leaves, and then dab them on the wound again. Infants lick their mothers' wounds, but adults do not usually lick the wounds of other adults.

People have long noticed that domestic dogs lick their wounds persistently, and that the wounds stay clean and free from infection. It might seem strange that a tongue used one moment for cleaning the anus can be used seconds later for cleaning an open wound, but dog saliva contains antimicrobials capable of killing bacteria such as *Staphylococcus*, *Escherichia coli*, and *Streptococcus*. The saliva of all mammals is an excellent disinfectant for their own wounds, and studies on rodents have showed that their saliva also contains epithelial and nerve growth factors that speed up the closure of wounds.[15]

Human saliva contains healing substances: mucins and fibronectins inactivate microbes by binding them; lactoferrins kill iron-dependent bacteria by taking their iron; peroxidases poison bacteria;

histatin is a strong antifungal agent; and one type of antibody, IgA, that is particularly common in saliva is active against viruses such as polio and influenza. Licking our own wounds is therefore a simple but effective way to enhance wound healing.[16]

Some animals even employ the skills of other species to help heal their wounds. Blue tangs (herbivorous group-living Caribbean coral fish) frequently sustain small cuts and abrasions yet seem not to get any infections. When seriously wounded, these fish, like many mammals, leave the group and stop feeding. They increase their visits to "cleaning stations," where they allow wrasse fish to feed off their dead and infected tissue. After the injury is completely covered by a scab, the coral fish resume normal feeding and reduce their visits to the cleaning stations. The help they get is beneficial, as even after deep wounding of subcutaneous tissue there are no visible signs of scarring.[17]

Pus is seriously underrated. Reindeer do not lick their wounds, but they do produce copious quantities of pus. The pros and cons of pus are currently much debated in human wound treatment. Since the advent of antibiotics that efficiently eliminate infection, we have come to see pus as a danger sign that should be immediately eradicated. Of course serious wounds should not be left unattended and open to infection, but in the right circumstances, with a healthy immune system and an appropriate diet, pus (like fever and diarrhea) may not be a bad sign.

## HERBAL HEALING

According to Chinese folklore, many centuries ago a farmer in the Yunnan district found a snake near his hut. Fearful for his life, he beat it senseless with a hoe and left it for dead. A few days later, the same snake returned. Again the farmer tried to kill it, but again it returned. After he had beaten it a third time, the farmer followed the severely wounded snake as it crawled into a clump of weeds, started feeding on them, and thereby rapidly cured the worst of its injuries. The plant in the story was *Panax notoginseng*, which now forms the main ingredient in the herbal formulation "Yunnan bai yao," a white powder that cauterizes cuts and stems external bleeding immediately.

It was standard issue to North Vietnamese soldiers during the Vietnam War, for use when they were wounded far from conventional medical treatment.[18]

The snake legend bears all the hallmarks of a tale designed to communicate a message rather than facts. As a rule, snakes do not eat plants — although this fact does not eliminate the possibility that they may use plants in medicinal ways. As in all good legends, our hero returns *three times* to make his point. And it is no coincidence that our hero is a snake. Snakes are particularly adept at healing serious injuries, which no doubt explains why they are so common in medicinal symbolism and folklore. Snake expert Harry Greene has seen snakes recover after being cut almost in half. He suggests that their phenomenal wound-healing ability may relate to their being cold-blooded — which means that they do not go into shock as warm-blooded animals do. Perhaps this story served to remind people of the efficacy of a local wound-healing plant. Similar tales have turned up in other parts of the world — as we shall hear later.

A European legend also tells of animals using herbal remedies to speed the healing of wounds. It is said that peasants in the Neydharting area of Austria, who have traditionally drunk and bathed in the local moor waters as a cure-all, learned its properties from watching wild animals. In memory of this legend, the coat of arms of a nearby village depicts a wounded buck bathing in the moor waters. Wildlife in the area is reputed to prefer to drink the waters of the moor even though clear spring water is available in a lake nearby, and folklore asserts that it was common to see a stag wounded in the hunt drag himself to the moor, sometimes across vast distances. There he would immerse his open wounds in the black muddy waters until he invariably recovered. The mud of the moor, used by contemporary European veterinarians in the treatment of wounds, contains over three hundred bioactive herbs, numerous trace elements, organic substances, sulfur, and many antimicrobials, vitamins, and hormones.[19]

A plant-eater's diet may fortuitously contribute to the efficient healing we have seen in the wild. Certainly some plants are capable of enhancing wound healing. Asiaticoside, for example, derived from the plant *Centella asiatica*, hastens healing by increasing tissue production.[20] Antioxidants such as flavonoids and lignans are important as well in wound healing and are common in plants eaten by wild animals.

Primate researchers in the field often hear tales of lemurs, howler monkeys, and other primates applying medicinal plants to wounds. Certainly ethologists have seen both monkeys and apes use leaves to wipe surface blood from wounds. In 1955 a natural historian collecting museum samples came across a newly captured gibbon with a severely swollen wound. When the wound was lanced, it contained masticated leaves of a medicinal plant natives used to treat wounds.[21] The collectors assumed that the gibbon had treated himself with the leaves and the wound had healed over them. However, because gibbons are often kept as pets in this part of the world, we cannot rule out the possibility that he had received medical attention by humans.

Dian Fossey reports an outbreak of violence between two groups of gorillas that left blood-splattered vegetation and one adult male slumped motionless under a tree. His face contorted in pain, he reached out for a few strands of *Galium*. A strange time to be browsing for food? *Galium* plants in this region are rich in astringent tannins that potentially help wound closure and fight infection. He began to lick the forefinger of his right hand, passing it repeatedly from his injured clavicle to his mouth, putting saliva (and perhaps plant extracts) on the wound.[22]

Primates certainly have the ability to treat their own wounds. One captive capuchin monkey called Alice was wounded by other monkeys so badly that she required stitches near her vagina. She groomed the area intensively for days, which was not unusual in itself, but then she took a stick, chewed one end to make a brush, and used it to apply syrup (supplied as food) to the wound area. She did not use tools to groom any other part of her body, nor did she apply any substance other than syrup. This strong sugar solution is an excellent ointment for wounds — soothing and antibacterial (strong sugar solutions literally explode bacterial cells). A natural correlate, honey, is commonly used in traditional medicine for the same purpose and is recommended by Western medics as a first-aid treatment for wounds. Alice has never applied syrup to her body at any other time — only when wounded. A few years later, Alice's infant received a lethal wound to the head from other monkeys. Alice not only licked and groomed the wound, but made a tool and applied syrup to the wound, as she had done to her own wound long before.[23]

It has been suggested by herbalists and natural historians that many nonprimate species treat their wounds. Wounded elk, moose, bears, and caribou roll in clay; bears and deer rub on resinous trees; wounded cattle and deer roll in sphagnum moss; deer stroll out into the salty sea; and an assortment of wounded animals are said to dip into cold water to staunch bleeding or numb discomfort. Herbalist Raymond Dextreit tells of a sea resort in the Siberian forests of Oussouri where the curative properties of earth were discovered through observations of wounded wild pigs, roe deer, red deer, and other animals who came to wallow in the mud. Anecdotal observations of wounded animals have reputedly led to other medicinal discoveries. When wounded deer, for example, were seen to rub their wounds on the sweet gum tree, Indians discovered that the resin had antiseptic properties.[24]

Elephants are frequently seen to cover their wounds, and Joyce Poole of the Kenya Wildlife Society is convinced that elephants dislike the sight of blood. Iain and Oria Douglas-Hamilton, who made the first systematic study of elephant behavior in the wild, report an incident in which a hunter shot a bull elephant in a small group. As the bull lay dying, the other elephants packed the wound with mud.[25]

Asian elephants also put mud on open wounds. One hunter watched a female elephant squirt mud all over her back in what he considered a rather strange manner. When he looked more closely, he saw she had a large wound on her flank from a tiger scratch. Captive elephants often get sores under their leg chains. One elephant at Washington Park Zoo in Oregon picked up the only plant that was available to her, a lettuce leaf left over from her dinner, and rubbed it purposefully over the wound with her trunk as if trying to soothe her discomfort.[26]

## PLASTER CASTS

Animals have even been reported to make natural plaster casts. Maurice Mességué claims that a Pyrenean goat with a broken leg has been known to make itself a plaster of clay and grasses, using its mouth. I might have ignored this unsubstantiated anecdote except for finding a similar description by biologist Lyall Watson: "I know of a record of a mountain goat with an injured leg that actually made itself a poultice of lichen and clay and applied it to the wound." My curiosity aroused, I pursued the modern myth but could not determine its origin.[27]

I did find an account of an injured woodcock written in 1903 by William Long: "At first he took soft clay in his bill from the edge of the water and seemed to be smearing it on one leg near the knee. Then he fluttered away on one foot for a distance and seemed to be pulling tiny roots and fibres of grass, which he worked into the clay he had already smeared on his leg. Again he took more clay and plastered it over the fibres, putting on more and more till I could plainly see the enlargement." Long then offered his own anthropomorphic interpretation of events: "The woodcock had broken a leg, and had deliberately put it into a clay cast to hold the broken bones in place until they should knit together again." An acrimonious public battle ensued, in which President Theodore Roosevelt called Long a "nature faker." Theodore Wood, a leading British natural historian of the time, stepped into the fray to say that many times sportsmen and natural historians had seen snipe with legs broken by shot, that bound them with feathers, leaves, and sticky substances. However, as neither Wood nor Long was a scientist, their observations (not helped by unsubstantiated interpretations) were easily dismissed.[28]

I am loath to suggest more research on this topic for fear that some misguided investigator will break the bones of a variety of animal victims in order to explore their responses! However, there is obviously a need to *observe* how animals behave when accidentally injured in the wild. Elephants certainly do apply clay to injuries, and other species may do so as well — but it is probably because of the immediate soothing or cooling effects of the clay, not its immobilizing properties.

## INVALID CARE

Just as humans find touch a valuable therapeutic tool, so too do injured animals. A seriously hurt young chimpanzee will calm down under the relaxing touch of its mother. At Gombe, when Mandy's three-year-old infant appeared with a torn and bleeding arm, the bone protruding, her pain was obvious; she held her head rigid with eyes open, glazed, and staring. As Mandy groomed her, she relaxed and her eyes closed. However, care for the sick is rare among unrelated chimpanzees at Gombe and only family members wipe the wounds of others. When one female, Fifi, had an infected gash in her head, those individuals to whom she presented the wound for grooming seemed fearful and moved away. When nine-year-old Freud broke his ankle, he was only able to move very slowly. Both his mother and his younger brother waited for him, but it was his brother who made the most fuss, whimpering to make their mother wait, grooming and staring at his injured brother's ankle.[29]

A week after a female mountain gorilla was badly injured in an intergroup fight, her bite wounds were draining badly and, had it not been for her five-year-old daughter, would have taken far longer to heal than they did. The daughter licked and probed stubbornly at the bite injuries on the back of her mother's shoulders, neck, and head until all had healed six weeks after their infliction. Gorillas can anticipate injury, too. Young gorillas whose wrists are caught in poachers' wire snares are rescued by their group silverback, who slides his canine teeth under the wire and lifts it off their hands.[30]

Invalid care is rare among social mammals, but dwarf mongooses in the Taru Desert of Kenya are one exception. Anne Rasa watched wild dwarf mongooses nurse a seriously injured female member of their group. She had lost the skin from her lower abdomen and inner thigh, and was stiff with discomfort, but the group stayed huddled round her, grooming her and bringing food. For six days, while the injured mongoose was unable to walk, the others restricted their normal foraging behavior to nurse her.[31]

Among those most famed for their nursing skills are elephants, who will pull spears or darts out of each other, help others to their feet, and even rescue one another in hunting raids. Unusually, altruism in elephants is not limited to relatives, or even to their own species, demonstrating a highly developed awareness of the needs of others.[32]

## BACK TO SNAKES . . .

Humans, often fearful of snakes themselves, have long been fascinated by how other animals deal with them. Folklore is rich with stories of animals arming themselves against poisonous snakebites. Hedgehogs and weasels are said to roll in the leaves of plantain (*Plantago* sp.) — a well-known vulnerary (wound-healing) herb. Herbologist Edward E. Shook writes that, when bitten, the mongoose "dashes into the jungle, finds a herb, eats it and rubs the poisoned parts in its juice, then returns to kill with immunity to the snake poison." This has been seen, he claims, by "thousands of witnesses" and "no mongoose has ever died of snakebite." No scientist, however, has documented this behavior or its unlikely interpretation — although some species, such as ground squirrels, do have a remarkable resistance to snake venom, and approximately eight hundred plants are currently known to offer antivenom properties.[33]

Another piece of contemporary folklore concerning snakebites resembles the ancient Chinese legend related earlier. An elderly Appalachian hill farmer, Harley Carpenter, recounts how two men came across two snakes tangled up in the road, fighting. One was a big black snake, the other a rattlesnake. The rattlesnake bit the black snake, which disappeared for a short while and then came back to continue the fight. Each time it was bitten, the black snake would go off and eat a clump of weeds. After watching this three times — yes, three times — one of the men reached down and pulled up the weed. When the black snake went back for more, it was of course unable to find the weed and died of its rattlesnake bite. Carpenter says that this observation led local people to assume that the weed might be an effective antivenom.[34] Like the Chinese legend, this piece of folklore is most likely an aide-mémoir among people who rely on natural medicines for their health.

We are not alone in our distrust of snakes. Many other animals become highly anxious in the presence of snakes and perform various behaviors to warn others. When kangaroo rats see a snake they drum their feet on the ground, and the great gerbils of Uzbekistan and Turkmenistan whistle while thumping with their feet. Vervet monkeys give alarm calls and cluster around snakes. Chimpanzees at

Gombe also raise a fuss when they see a snake. Once, when Gremlin was carrying her son Gimble along a trail, she saw a small snake ahead. Carefully she pushed Gimble off her back and kept him behind her as she shook branches at the snake until it glided away.

Research on human phobias shows that we have a predisposition to learn a fear of snakes — a trait that must have stood us in good stead in our more rural past. Californian ground squirrels, though, have been perplexing scientists for many years: they will *deliberately* taunt rattlesnakes to the point of being bitten, sometimes having to pull the fangs out of their little furry bodies before taunting the snakes again![35]

It seems ground squirrels are resistant to the venom, but this does not explain why they should be so brazen. Current explanations revolve around the idea that the squirrels are assessing a dangerous opponent by taunting. But being bitten seems to negate such an assessment. Self-medication is another possibility: just as plant toxins can be beneficial in small doses, or when the recipient has evolved a mechanism for dealing with them, so snake venoms can be medicinal or can stimulate beneficial or pleasurable reactions. Research on snake venom shows that it is capable of stimulating the immune system generally and thereby slowing the spread of cancer.[36] This phenomenon is poorly understood, but we cannot rule out the possible health *benefits* of being bitten — among other reasons, as an explanation for the ground squirrels' flagrant taunting of snakes.

The assumption that animals deal with injury merely by avoiding unnecessary risk and relying on their immune systems to do the rest has severely limited our understanding of natural pain management and wound treatment.

# 8

❦ ❦ ❦ ❦

# MITES, BITES, AND ITCHES

> Great fleas have little fleas upon their backs to bite 'em,
> And little fleas have lesser fleas, and so ad infinitum.
> And the great fleas themselves, in turn, have greater fleas to go on;
> While these again have greater still, and greater still, and so on.
>
> — Augustus De Morgan, 1850

"NOTHING CAN STOP the omnipresent mosquitoes, bottlas, and sand flies from feasting on any inch of your flesh that is left available," wrote wildlife biologist Alan Rabinowitz on the trials and tribulations of studying jaguars in the rain forests of Belize. "If it's been an unlucky day, you may have been bitten by a mosquito carrying the malaria protozoan or yellow fever virus, or by a sand fly carrying *Leishmania,* a parasitic protozoan. Or a mosquito may have deposited a botfly larva, which will burrow beneath your skin. You won't know it until weeks later, when it starts to eat its way out."[1]

For pests that can pierce this protective layer of skin, the bountiful reward is nutrient-rich blood, body tissues, and a safe nutritious place to lay their eggs; for fungi and bacteria, it is a continuous supply of body secretions on which to feed. Mammals and birds have particular problems in this regard; although fur and feathers may be great for insulating and waterproofing the skin, they also provide warm, moist cover. More than two thousand species of fleas alone infest birds and mammals. And beyond those are lice, mites, ticks, and an army of biting and parasitic insects.

These "mites and bites" are more than a minor irritation. Left unchecked, tiny blood feeders can kill. A horse can lose up to 0.5 liters of blood a day to blood-sucking flies, and a mere six engorging ticks can impair the chances of survival for wild impala or slender gazelles grazing on the plains of Africa.[2] Lice-infested cattle grow more slowly, and lice-infested swallows produce smaller chicks and have fewer clutches in a year than noninfested swallows.[3] Biting insects also spread the world's great plagues: malaria, bubonic plague, yellow fever, sleeping sickness, dengue, and filariasis, causing over a million human deaths each year. Bubonic plague, which ravaged Europe in the Middle Ages and still threatens the health of people in Asia, South America, and Africa, is spread by the bites of fleas living in the fur of black rats and other mammals. And malaria, estimated to have killed half of all humans who have ever lived, is spread by the blood-sucking bites of mosquitoes. Initial hopes that DDT would control mosquitoes were dashed as they became resistant and spread ever more widely.

Even bites that do not carry disease can leave the skin open to attack from fungal and bacterial infection, and mere itchiness can distract an animal from getting on with other tasks necessary for survival. As these mites, bites, and itches are potentially so damaging, selection has favored animals that have found ways of dealing with them.

## DODGING AND DIVING

The tenacious ability of biting insects to find exposed skin is well known — and not just in the tropics. Scotland and Alaska can offer as many biting insects as Africa or India. How do wild animals exposed to the ravages of these insects twenty-four hours a day manage to survive the constant attacks on their skin?

First, animals do all they can to avoid being bitten. Biting insects often locate prey by following the trails of carbon dioxide they exhale, and when closer, by homing in on body warmth. North American caribou confuse mosquitoes, gadflies, and biting midges searching for warm bodies by huddling together on small patches of ice high up on the tundra. Other species also exploit localized weather conditions. When flies are being particularly bothersome, feral horses move to windy hills where, although the grazing may be poorer, flies find it difficult to land.

Even with their thick hides, elephants, buffaloes, and rhinoceroses are vulnerable to specially adapted bloodsuckers, and they roll in thick mud for extra protection. Humans find this approach useful too. The Jarawa people of the Andaman Islands of India coat themselves in clay to protect against the bites of mosquitoes; when collecting honey, they mix the clay with plant extracts to protect themselves from bee stings.[4] Another form of protection is immersion. Tigers and jaguars, for example, can spend most of their day sitting or lying in water, with only their heads above the surface. As well as avoiding the ravages of numerous biting insects, they stay cool while keeping an eye on potential prey.

Some animals even make their own fly swats. Forest-dwelling Malay elephants carry bunches of palm leaves in their trunks to flick away flies. The Asian elephant in Nepal goes one stage further, not only picking and carrying a fly switch, but modifying it for the job — stripping off surplus leaves and twigs to obtain just the right length and flexibility.[5] Chimpanzees also swat flies with leafy twigs.

Fidgeting, twitching, and constantly moving are effective ways of avoiding bites. A horse bothered by blood-sucking flies shakes its head, tosses its long mane and tail, stamps its feet, gallops around, rolls in mud, twitches its skin, bunches together with other horses, or stands by the edge of forest fires to smoke the flies away. A similarly harassed camel runs into the herd and rubs against other camels. Elephants are able to twitch their skin with such force that tiny pests are crushed in the skin's folds. Large, majestic herons stamp and peck at mosquitoes around their feet up to three thousand times an hour, but it is worth the effort because this behavior can successfully prevent more than 80 percent of the mosquitoes from feeding on the herons' blood. Although constant movements consume valuable energy, the cost of *not* moving can be higher still. If exposed to many mosquitoes and restrained from any motion, a rat will die within a very short time, presumably from loss of blood.[6]

Fleas, mites, lice, and ticks (so-called ectoparasites) are often passed to infants from the mother or bedding soon after birth, and thus may be harder to avoid than biting and parasitic flies. Even so, animals have evolved ways of limiting their exposure. Both sea-birds and swallows simply abandon nesting colonies and move on when infestation levels become too high. Veterinarian Benjamin Hart thinks the need to avoid the buildup of parasites may even have contributed to the evolution of migration, as whole populations moved away from heavily infested areas and returned only after parasite levels had fallen.

European badgers have a more complicated arrangement for avoiding ectoparasites. Their large underground chambers, collectively known as a sett, are dug over several decades by generations of badgers. One sett in southern England is known to be at least two hundred years old. Such an investment is not abandoned lightly, so badgers move around within the sett. They change nest chambers as numbers of ectoparasites build up, each badger careful not to sleep in a chamber used the previous night by another badger.[7]

## PICKING AND PLUCKING

Birds and mammals spend a great deal of time attending to their skin, fur, or feathers. African antelopes groom themselves a thousand times a day and scratch at least another thousand times. Great tits with young chicks spend almost a third of their normal sleeping time removing bloodsucking hen fleas from their nests at night so that they do not lose feeding time during the day. Grooming is highly effective at reducing the numbers of lice and fleas, sometimes as much as sixty-fold.[8] Even so, there are inevitably awkward places that hands, paws, claws, mouth, or beak cannot reach. Consequently, some animals groom each other reciprocally. This so-called allo-grooming is particularly common in species that form social hierarchies and is thought to play a strong role in cementing alliances.

Most primates spend a large portion of each day grooming one another: carefully parting the hair, picking out ectoparasites, and occasionally eating them. This attention explains why free-living primates typically have so few ectoparasites. Chimpanzees at Gombe remove ticks almost as soon as they attach, sometimes even preempting a tick attack. Once Gremlin was walking with her youngster,

Gimble, and others along a trail, approaching a patch of tall grasses. Another adult female, Melissa, went through the grass, but when Gremlin got there, she prevented Gimble from following Melissa's lead and pushed him behind her. She hit the clumps of grass several times and herded him carefully around. On inspection the grass was found to be infested with hundreds of minute ticks.[9]

Allogrooming is so important that at least one species of mouse uses it as payment for sexual services. The female wood mouse will allow mating to proceed only after a male has adequately groomed her. In a similar vein, allopreening is a marital bonus for eudyptid penguins. Mated pairs groom each other and consequently have far fewer ticks than solitary birds that can only preen themselves.[10]

Even with the help of conspecifics, there are still nooks and crannies that are difficult to groom. In these cases mutually beneficial relationships can evolve between species. In the forests of Brazil, black caracaras (long-legged birds of prey) clean ticks from large bare-skinned tapirs, and pale-winged trumpeters clean gray brocket deer of insects and ticks. In a forest where visibility is poor, they call out to attract one another for this crucial exchange.

The oceans too are abundant with these food-for-cleaning swaps. Biologists even suggest that many marine animals would not be able to survive the ravages of ectoparasites, bacteria, and fungi were it not for the attentions of cleaners. In the Indo-Pacific coral reefs, the dangerous moray eel allows small cleaner wrasse fish to swim in and out of its mouth to collect ectoparasites and diseased tissue. In recent trials wrasse fish have proved effective cleaners of farmed salmon — an observation that might lessen the industry's reliance on harmful chemicals.[11]

Not all cleaning relationships are mutually beneficial. Some cleaners may be unwilling slaves. The Eastern screech owl feeds mainly on insects but brings live Texan blindsnakes unharmed to its treetop nest, where they feed on the larvae of scavenging or parasitic insects that are bothering the young owls. Nests that contain blindsnakes are more successful at rearing young than those that do not, so the benefit to the owls is obvious. But while blindsnakes may get a few meals, they are often found dead in the nest, suggesting that they do not always share in the benefits.[12]

## RUBBING AND ROLLING

It is evident, then, that animals adopt a variety of strategies for physically reducing or removing skin pests. What scientists are now finding is that sometimes animals also use nature's pharmacy to help — they have *medicinal* strategies as well. For example, capuchin monkeys living in the forests of South America rub into their fur a variety of natural substances that repel pests, soothe sores, or deaden itches. Among a group of capuchins in Costa Rica studied by Mary Baker, fur rubbing occurs most commonly in the wet season. During each bout a monkey spends about six minutes in "frenzied and highly energetic" fur rubbing; it bites and rolls a plant between its hands while applying the plant-saliva mixture over its entire skin. The result is a mass of drooling, wet monkeys with bits of fruit pulp, juice, and broken leaves stuck to their fur, squirming and rolling over and around one another. Baker says, "They really get into it, drooling like crazy, spit flying everywhere." The plants they use include the stems of *Clematis* and leaves of *Piper* plants. In laboratory experiments *Clematis* kills bacteria and deadens pain, while *Piper* (the family of peppers and chillies) contains numerous volatile compounds with insecticidal and pain-numbing properties.

Citrus fruits are especially favored by fur-rubbing capuchins. Taking only the rind, they usually abrade it by biting or pounding it on tree branches. Other times they break open a citrus fruit and hug it to their chests and stomachs, at the same time digging into it and rubbing it hard into their fur. In captivity, capuchins rub *anything* smelling of citrus onto the skin, including lemon-scented soap. Citrus is both pungent and stimulating, containing volatile oils and flavone glycocides that have analgesic, insecticidal, and antimicrobial properties. When Baker tried rubbing her own skin with citrus peel in the field, she found that her mosquito bites stopped itching.[13]

Capuchins also rub the fuzzy seedpods of *Sloanea terniflora* over their bodies until all the scratchy hairs are worn off — at which stage the monkeys find fresh pods and start again. Little is known about this plant's chemical properties, but it seems likely that the pods are used simply for scratching. European badgers employ thistles in a

Wedge-capped capuchin monkeys rub toxic millipedes like this one into their fur. The millipede's secretions are antimicrobial and repellent to insects. *Ximena Valderrama (monkey)/Thomas Eisner (millipede)*

similar way — vigorously rubbing the prickly plants over their bodies, apparently with great delight.[14]

Plants are not the only part of nature's pharmacy that capuchins utilize. In the llanos of central Venezuela, wedge-capped capuchins are exposed to intensive attack from insects, especially mosquitoes, during wet-season floods. When this happens they take advantage of millipedes that secrete toxic benzoquinones, which are both repellent to insects and antimicrobial. Upon finding one of these large (as long as 8 centimeters) millipedes, a capuchin gets it to release its defensive toxins by rubbing it and rolling over it, intermittently taking it into its mouth without eating and slowly withdrawing it again. During mouthing the monkey drools copiously and its eyes glaze over. One millipede can be shared by several capuchins; those that do not have one of their own rub against others that are already covered in secretion. The result is the anticipated writhing cluster of drooling monkeys.[15]

In captivity, capuchins have showed themselves capable of curing skin conditions. One captive capuchin was prone to necrotic skin infections that required repeated veterinary treatment. When he was given daily access to tobacco leaves (containing the potentially toxic alkaloid nicotine, which has a strong smell and bitter taste) he effectively self-medicated his skin condition, which not only disappeared but never returned.[16]

Many other mammals rub bioactive compounds into their fur. In Panama, white-nosed coatis, relatives of raccoons, rub resin from the *Trattinnickia aspera* tree into their coats. This tree is a member of the Burseraceae family, which ordinarily secretes resins with a smell similar to turpentine, but this species (known only in Panama) also has a camphor, or menthol-like, odor. Coatis travel a long way to get to their special grooming trees, and they certainly seem to enjoy fur rubbing. As they approach, some individuals get so excited that they break into a run, reaching the tree before the others. They rupture resin ducts, digging at wounds in the tree base to allow more resin to flow. Soon the trees become ringed with deep menthol-smelling wallows. Quickly moving their paws over their entire bodies, including their tails and faces, the coatis vigorously groom the oozing resin into their fur and skin. In their haste, they pile on top of one another and rub resin into each other as well. The frantic scene is over in minutes.[17]

Camphorated oil is used by herbalists around the world against ectoparasites, and this particular camphor-smelling resin is employed by the local Guaymi people for medicinal purposes. Chemists at Cornell University have identified triterpenes, amyrin, selinene, and sesquiterpene lactones in the resin. The last of these are known to be repellent to fleas, lice, and ticks, as well as to biting insects such as mosquitoes.[18]

Black, brown, and kodiak bears reputedly make their own herbal paste. They dig up osha roots (*Ligusticum wallichii* and *porteri*), chew the root, and frenetically rub the root-saliva mixture into their fur.[19] Chewing the root presumably releases the active constituents and mixes them with saliva for easy application. According to legend, the bear taught Native Americans how to use the root they call "bear medicine" as a topical anesthetic and antibacterial — another instance where observation of wild animals has reputedly led to discovery of a medicine for humans. Unfortunately, the popularity of the root in the global market for herbal products has left the plant endangered in the wild.

Osha root is aromatic, containing volatile and fixed oils, a lactone glycoside, an alkaloid, phytosterols, saponins, ferulic acid, phthalides, and monoterpenes. In Native American herbal medicine it is used topically for external skin conditions and bruises. It is a strong analgesic for the throat and a mild antiviral, so chewing prior to fur rubbing may provide additional medicinal effects. Captive bears reputedly show the same fur-rubbing behavior when provided with *Ligusticum* roots or similar pungent plants. And bears may have other skin rubs: in Quebec, natural historians have seen bears rub on resinous fir trees, becoming covered from head to toe in a sticky resin used by local Indians to repel bothersome black flies.[20]

Cat owners know that their pets love to roll in catnip (*Nepeta cataria*). They often salivate and appear transfixed by the pleasure (see Chapter 10). Still, there may be more to this scene than fun. An active ingredient in catnip, nepetalactone, is so effective at repelling pests that it deters even the seemingly unchallengeable cockroach. Joy Adamson often saw her orphaned leopard roll in an African relative of catnip, *Leonotis nepetifolia*, otherwise known as catnip leaf. Although she supposed at the time that her leopard rolled in the pungent plant to disguise her own scent while hunting, many of the plant's active ingredients are insecticidal or pesticidal.[21]

One animal that seems to have particular difficulty managing its skin is the European hedgehog. Beneath its sharp spines, it is vulnerable to fungal ringworm, mites, fleas, and ticks. Normal grooming is difficult for obvious reasons, which may explain why hedgehogs anoint themselves with a variety of pungent substances: mints, tobacco, oil, and fermenting fruit, to name only a few. When self-anointing, a hedgehog will salivate profusely until it produces a large quantity of frothy saliva; it will then flick the saliva over its body with its long tongue, contorting itself to reach all parts of its body. Folklore tells us that hedgehogs anoint themselves to dissuade predators, but that would seem an unnecessary effort, considering the seriously unattractive prickles that have evolved for that purpose. On the other hand, the need to deal with those inaccessible mites, bites, sores, and itches provides ample justification for self-anointing.

Obviously, much mammalian fur rubbing could be preventive against skin pests, but rubbing with astringent and analgesic substances may also provide rapid relief from itches and sores. The skin rubs may therefore afford both preventive and curative effects. The dribbling and drooling during fur rubbing may be particularly significant. Salivation acts as a medium to spread a substance through the fur, but differs from chewing or spitting. It is a reflex of the central nervous system — an involuntary response. And this might provide an exciting clue to how animals select their herbal remedies.

Research on humans finds that salivation is a direct indication of the insecticidal properties of certain secondary compounds. When assessing the insect-killing actions of isobutylamides (also found in the *Piper* plants used by capuchin monkeys), the phytochemist Francis Brinker found two effects: one was that people salivated, the other was that these plant compounds produced a burning sensation on contact with the skin. Most important, the amount of salivation was proportional to the effectiveness of the compounds against insects. If human salivation is a valid indicator of insecticidal activity in compounds from the *Piper* family, salivation in mammals may be a general indicator of such properties in the plants they use. In addition, we should not forget the medicinal (antiseptic) nature of saliva.[22]

Birds too rub bioactive substances deep into their feathers and skin. Over two hundred species rub live ants into their feathers — so-called anting. Most commonly, a bird crunches an ant in its bill and rubs the ant frenetically through its plumage. Other times a bird might entice ants to crawl through its plumage, by crouching or lying on an anthill with spread wings and tail. Although anting was first recorded in birds, ant nests are also sought by squirrels, cats, and monkeys, who roll in them with apparent delight.[23]

The ants preferred by birds and mammals are those that secrete toxic fluids such as formic acid, a bitter and pungent substance with properties similar to those of the *plants* chosen by mammals for skin care. In laboratory tests, formic acid kills chewing lice, and the vapor alone can kill lice and feather mites. One extremely patient (and dextrous) scientist managed to catch four wild steppe pipits in the process of anting, and confirmed that their feather mites were dying.[24]

Formic acid is also analgesic, which has led some scientists to suggest that birds and mammals use it to allay the discomfort of molting. This may be true, but anting is not seen *only* during molting. Furthermore, Dale Clayton has found that molting itself substantially reduces accumulated microorganisms and ectoparasites.

The medicinal benefits of formic acid have not escaped the attention of beekeepers around the world, who use it to control parasitic mites of honey bees — particularly the verroa and tracheal mites.[25] In addition to formic acid, ants secrete a range of other complex substances, such as auxins and beta-hydroxyl fatty acids that can kill fungi and bacteria, suggesting that anting is an all-round skin-care strategy.[26]

Dale Clayton and Jennifer Vernon watched a common grackle rub a lime fruit into its feathers. For twenty minutes the bird tried to balance itself on top of the discarded lime, hammering repeatedly with downward blows, then preening itself with bits of lime in its bill. The bird seemed unusually preoccupied and frenzied; when it had finished, pieces of lime pulp and outer rind were gouged or missing. Since then, many other scientists have seen grackles rub limes and lemons into their feathers and skin. Citrus fruits have two possible mechanisms of medicinal action: one is direct contact with the skin, the other is the action of the vapors. In traditional medicine, lemon juice is topically applied to kill ringworm, as the constituents are capable of killing fungi and bacteria. A monoterpene, *D*-limonene, present at concentrations of 98 percent in the peel oil of citrus fruits, is toxic to a wide variety of arthropods, such as lice and fleas. When Clayton and Vernon tested the effect of lime peel on bird lice in the laboratory, they found that nine hours' exposure to the vapor was enough to kill the lice.[27] Clayton suggests that the vapor may act as a deadly gas trapped in the birds' feathers, in the same way that trapped air acts as insulation to keep the birds warm.

Just as herbalists use citrus peel to repel fleas, and aromatherapists use citrus oil as a stimulant, animals that rub citrus on their skin can potentially medicate against ectoparasites, fungal and bacterial infections, and feel an exhilarating rush all in one fell swoop! Perhaps its popularity as a skin rub is not so surprising after all. What is surprising is the wide use of citrus by both birds and mammals in the Americas, where it is a relatively recent introduction from its native Asia.

In addition to rubbing insects and citrus fruits into their skin, birds rub an array of aromatic leaves, flowers, and other substances through their feathers. Most of these substances have antimicrobial or insecticidal properties. The frenzied nature with which substances are rubbed into fur or feathers may be relevant. Mary Baker comments that when she shows people her video of capuchins fur rubbing, they assume that she has it on fast-forward! A degree of frenzy is not too surprising, in that wild animals are likely to be covered with numerous tiny insect bites, sores, and minor irritations. Try rubbing citrus peel into sores and itchy skin. It'll make you feel a bit frenzied, too!

Animals have several other ways of keeping their skin healthy. Birds bathe in dust, rolling around and shaking it into their feathers and skin; the dust soaks up excess feather oils and dries the skin surface, making it less hospitable to microorganisms. Fine soils also contain tiny, sharp particles that cut the exoskeleton (outer skin) of small ectoparasites. For these reasons dust is recommended by the Nature Society for controlling blowfly larvae and mites on captive birds. Mammals too, particularly bare-skinned elephants, like to bathe in dust.

Many animals sun themselves. Birds lift their wings to expose the undersides to sunlight. Clayton hypothesizes that dusting and sunning may play a role in microbial defense by making the plumage too dry to support bacteria.[28] Gorillas sunbathe in small glades of the forest; rabbits lie in the sun, exposing their white underbellies, contorting their upper bodies so as to maintain watch for potential predators. European badgers even bring their bedding up to the surface for a day or two of "airing" before returning it to their underground nesting chambers. Sunlight, particularly ultraviolet radiation, is an effective method of sanitation, killing bacteria and viruses. Combined with desiccation, it could have a significant effect on emerging larvae of mites and lice, and skin conditions.

Salt is another widely used remedy for skin problems: its osmotic properties burst microorganisms and ectoparasite larvae. Mangy camels not only feed on salty plants such as *Sueda monoica,* but roll in salt; camel herdsmen regularly take them to natural salt pans for this purpose. They find salt an effective cure for mange, and the local people copy the camels and rub salt on their skin to cure scabies (a mite infestation).[29]

Zookeepers have long known that captive capuchin monkeys urinate on their palms and wipe the urine over exposed parts of their skin, but as this behavior had not been reported in the wild, it was assumed to be an aberration of captivity. More recently, though, urine washing has been seen in wild moustached tamarins in South America. They seem to wash mostly at midday, when temperature is at a maximum and humidity at a minimum, to keep themselves cool. (As urine has a higher osmotic potential than water, it creates an even greater cooling effect than water alone.)[30]

It may be that urine washing has a pharmaceutical function as well. Urine from healthy individuals is sterile and antiseptic as well as cooling, and has traditionally been used in emergency medicine in the treatment of wounds, blisters, and chilblains. The American Indians of the Northwest Territories used urine for regular skin care, washing themselves all over with their urine each morning.[31] Urea (the main component of urine) is still used as an antibacterial and antifungal agent by physicians in the West, and veterinarians commonly use urea ointment to treat infected wounds. Washing with fresh urine could therefore act as a cooling disinfectant.

## POWERFUL SMELLS

The nests of birds, and the burrows and dens of mammals, are breeding grounds for disease and excellent locations for ectoparasites: dark, moist, warm, and providing a regular supply of bodies on which to feed. It is here that young animals are exposed to ectoparasites for the first time. As mentioned earlier, some species move on to avoid the buildup of these parasites, others repeatedly use time, energy, and resources to build new nests. If there were a way that animals could reduce cumulative infestation *without* moving or building new nests, we would expect natural selection to have favored it. Nature's pharmacy abounds with secondary compounds "designed" to repel plant pests and fight disease. By integrating these into their nests or dens, animals might conceivably get some protection from pests and diseases.

It has long been known that many bird species weave fresh greenery into their twiggy nests. In the mid-1980s Peter Wimberger noticed that North American birds of prey do so *only* around the time of egg hatching, and are more likely to bring greenery to the nest if they are reusing old nests. As old nests contain more ectoparasites than new ones, it appeared that the birds might be bringing herbs to fumigate them during the critical time when newly hatched chicks are exposed to ectoparasites for the first time. Further development of this nest fumigation theory has come from studies of the European starling.

In preparation for nesting, male European starlings bring fresh green plants and weave them loosely into the nest, although once the eggs are laid they lose interest in this activity. In North America they preferentially select wild carrot (*Daucus carota*), yarrow (*Achillea millefolia*), agrimony (*Agrimonia paraflora*), elm-leaved and rough goldenrod (*Soldaigo* sp.), and fleabane (*Erigeron* sp.), even when they are not the most common plants nearby. Interestingly, old European herbals refer to wild carrot as "bird's-nest root," suggesting that birds have been lining their nests with this plant for a long time.[32] The most obvious characteristic of this selection of plants is that they are all highly aromatic. Furthermore, they contain more volatile oils, in greater concentrations, than aromatic plants close at hand that are not selected. In other words, they are the smelliest plants around.

When Larry Clark and Russell Mason removed the fresh plants from some starling nests, they found that chicks in these nests were infested with more mites than those in which the green plants were left undisturbed. More specifically, chicks in nests containing wild carrot had higher hemoglobin levels than those without, suggesting that they were losing less blood to bloodsucking mites.[33] Interestingly, humans also use wild carrot for skin complaints: in traditional herbal medicine, wild carrot leaves are used to treat chronic itching. In addition, wild carrot is of the same family as the *Ligusticum* root used by bears on their skin.

As a group, the plants preferred by starlings are capable of reducing ectoparasite infestation in ways that the ignored aromatic plants nearby are not. They contain monoterpenes and sesquiterpenes (such as myrcene, pinene, and limonene) that are harmful to bacteria, mites, and lice in the laboratory. The preferred plants are particularly effective against the harmful bacteria *Streptococcus aurealis, Staphylococcus epidermis,* and *Psuedomonas aeruginosa,* but not against the usually harmless *Escherichia coli.* Although they retard the hatching of louse eggs and the emergence of mite larvae, the preferred plants do not *kill* either adult lice or adult northern fowl mites.[34] Fleabane, as its name suggests, has been known by herbalists through the ages to be repellent to fleas.

It appears, then, that starlings are choosing the *best plants available* to fumigate their nests against both microorganisms and ectoparasites, and they have shown themselves capable of detecting these volatile plants with some degree of accuracy. They can discriminate between the concentrations and numbers of volatile compounds in different plants, which vary with location — and they choose the plants with the most complex aroma. Their ability to detect volatile oils varies seasonally, being most acute at the time of reproduction when fumigation of the nest is of peak importance. Starlings are able to discriminate between a preferred plant (wild carrot) and a less preferred plant (red dead nettle) in April, at the beginning of breeding season, but not outside the breeding season in September. Seasonal changes in hormones could influence the male's ability to detect these significant plant odors.[35]

Bringing greenery to the nest is part of the male's courtship display, which females use to assess the attractiveness of potential mates. The aromatic properties of the plants collected could be important in the female's choice. Certainly, males are not all equally skilled at choosing the right plants; a large element of learning is involved. Young inexperienced male starlings are far more cosmopolitan in their selection of green nest material, but by the second year most starlings are conforming to the plant chemical profiles selected by their elders.[36]

In addition to their value for fumigation, the plants starlings choose are commonly used by herbalists for skin problems such as ulcers, sores, and eczema. Thus they may be able to help with the *symptoms* of ectoparasite infestation — scabs, sores, and itches — just as fur rubbing seems to do. Recent research in Europe suggests this may well be the case. When comparing grass-lined nests with herb-lined nests in Europe (where starlings prefer goutweed, hogweed, yarrow, white willow, elder, and cow parsley), scientists found that the numbers of mites, lice, and fleas were indistinguishable. Are European herbs less effective than those selected by starlings in North America? It seems not, for although the ectoparasites are not harmed directly, nestlings in herb nests weigh more and are less anemic at fledging than those in grass nests. In addition, more yearlings from herb nests are seen the year after hatching. In other words, the herbs selected by starlings in Europe are improving the health of the chicks — "helping them to cope better with the harmful activities of ectoparasites" — perhaps by enhancing the immune system.[37] Starlings in different locations are therefore improving their fitness by using different plants with different medicinal effects but the same end result: healthier offspring.

Many other bird species medicate their nests. Hawks select fresh greenery from a limited range of plants — all of which effectively repel insects in laboratory tests. In India, house sparrows bring leaves of the margosa, or neem tree (*Azidirachta indica*), to their nests at breeding time. These leaves contain numerous secondary compounds, among them azadirachtin, a complex chemical with powerful insecticidal properties, and sitosterol, a natural insect repellent that also disrupts egg laying in ticks and other blood-sucking parasites. Neem has been used in India for centuries as a remedy for skin ailments and to protect clothing from insects. A recent outbreak of malaria in Calcutta gave an unexpected twist to the tales of bird self-medication. The biologist Sudhim Senegupta noticed that during a malaria outbreak Calcutta's sparrows switched to lining their nests with quinine-rich leaves from the krishnachura tree (*Caesalpinia pulcherrima*). This tree is uncommon in the area, suggesting that the birds were preferentially seeking it out. As quinine is active against

Male European starlings line the nest with fresh aromatic herbs that help young chicks survive the ravages of ectoparasites. *Helga Gwinner*

malaria, Senegupta proposed that the birds might be using the leaves to medicate against the disease.[38]

One final example of bird pharmacy involves a bird that nests in small cavities rather than in an interwoven matrix of plant material. Nuthatches do not bring fresh green nesting material when reusing the cavities. Instead, they daub antiseptic and insecticidal pine resin (rich in terpenes, such as camphor) around the entrance hole and rub insects on the cavity entrance.[39]

Mammals may use aromatic herbs in a similar way. In addition to airing their bedding, European badgers bring a variety of dry bedding material to the sett. Near birthing time they start to collect *fresh* green material such as bluebell and daffodil leaves, dog's mercury, hart's-tongue fronds, ground elder, and wild garlic for use as bedding. Like birds, the badgers concentrate on finding fresh green plant material at the time when the vulnerable young will arrive in the nest. Furthermore, the plants commonly selected by badgers have properties similar to those chosen by birds. Most are aromatic, antimicrobial, and insecticidal, and they are used by herbalists in the treatment of skin complaints.[40]

Wolves may not live in two-hundred-year-old subterranean chambers as badgers do, but they do rear their cubs in underground dens. In the boreal forests of North America, David Mech noticed that wolves preferred to den under black balsam trees, even though these trees were relatively rare in the woodland habitat. Black balsams secrete oleoresins and volatile oils, including the familiar insecticidal and antimicrobial camphene and limonene that give the tree its characteristic balsamic odor. Mech does not claim to know why the wolves prefer to den under balsam trees; he simply notes that they do.

As any pet owner knows, many animals are attracted to, and enjoy covering themselves in, strong smells — often to our displeasure. Dogs and wolves roll in stinking carcasses with great delight. David Maehr has been trying to use radio collars to track black bears in Florida, but the bears rub so frequently on smelly creosote-soaked wood that the collars soon become disabled by a thick layer of sticky tar. Male golden bees seek out and collect "fragrances" from orchids and other plants that contain no nutrients.[41] Historically we have explained this attraction to strong odors in terms of pheromonal communication, or scent disguise. However, it is also possible that because volatile oils interfere with bacterial respiration and are commonly detrimental or repellent to arthropods and insects, rubbing in or collecting smelly substances could reflect an adaptive preference for compounds that enhance health. It is no mere coincidence that our household disinfectants are scented with lemon and pine — odors we associate with freshness and cleanliness.

## SYSTEMIC CONTROL

Some animals appear to have evolved yet another way of utilizing nature's pharmacy to combat mites, bites, and itches. Modern systemic flea treatments work from the inside out by oozing through the skin. In a similar way, wild animals feeding on certain plants may gain incidental protection. Many animal species accumulate toxic secondary compounds, making their flesh pungent and unpalatable. The flesh of the thrush-sized, orange and black pitohui birds of New Guinea is reputed to make the tongue go numb! From their diet of berries and insects, they store in their skin and feathers a steroidal alkaloid, homobatrachotoxin (found elsewhere only in poison-dart frogs). Bent Poulson is convinced that the storage of toxins in the skin and feathers is useful for deterring ectoparasites as well as predators. In laboratory tests, lice show far higher mortality on pitohui feathers than on nontoxic feathers; not surprisingly, they avoid feeding or even resting on them. Diet may also affect the composition of preen gland secretions. Scientists at Cornell University have found steroidal alkaloids in both ants and the preen gland secretions of ant-eating tropical birds, suggesting that the birds' diet is influencing their preen gland composition. By ingesting secondary plant compounds via insects, birds could be incidentally concentrating insecticidal chemicals in these glands. Diet may therefore be an underestimated method of skin medication for wild birds.[42]

To combat mites, bites, and itches, we have seen that animals physically remove the tiny pests, or get others to do it for them. They use mud, dust, and sunshine to dry up skin oils. They use aromatic, analgesic, and astringent plants and toxic insect secretions on their skin, in their nests, or in their food. The plants and insects they select are capable of providing broad benefits — a characteristic that is highly suitable to a mammal or bird suffering a multifaceted attack on its skin.

Before World War II, most insect repellents used by humans were derived from volatile secondary plant compounds such as citronellol, camphor, and menthol — all commonly used as skin rubs by wild animals. Although they were effective, we found that they evaporated quickly from our bare skin. The chemical industry accordingly developed artificial insect repellents such as DEET that last a lot longer. Unfortunately, the strong toxicity of these artificial insecticides can cause unpleasant side effects, and pharmacists are now *returning* to volatile plant compounds for safer insect repellents.

Chemists in India have recently determined that volatile mint oils can provide humans with 85 percent protection against *Anopheles culicifacies*, the mosquito responsible for three-quarters of malaria transmission in northern India. And the volatile oils coumarin and piperonal (found in lavender and violets) have been found to be more repellent to the yellow fever mosquito than DEET. As head lice (actually mites) become resistant to chemical treatments, herbal treatments that rely on volatile oils remain effective, when combined with physical removal by grooming (brushing). In short, we are returning to the sustainable methods of skin care used by animals.[43]

# 9

❉ ❉ ❉ ❉

# RELUCTANT HOSTS,
# UNWELCOME GUESTS

Unbidden guests
Are often welcomest when they are gone.
— William Shakespeare, 1591

UNABLE TO MANUFACTURE their own food, internal parasites feed, breed, and develop inside the bodies of other species. These freeloaders enter via the mouth, eyes, or insect bites, or force themselves directly through the skin. Once inside, they often pass through several different forms in various parts of the body, before producing eggs or larvae that pass out of the body and into the path of another unwitting host. In the process, they can wreak havoc.

As a species, we humans are not winning our ongoing battle with parasites. Amebic dysentery is a global health problem caused by a tiny single-celled parasite and spread by fecal contamination of food or water. It affects 40 million people each year and kills forty thousand. Chagas' disease, resulting from infection by a protozoal parasite, puts 100 million people at risk in South America, along with 150 species of wild or domesticated animals. Once the Chagas parasites lodge in the heart, death comes rapidly from cardiac failure. Malarial fever, triggered by a blood parasite and spread by mosquitoes, puts 40 percent of the world's population at risk at any one time. Alarmingly, many of these parasites are gaining resistance to modern chemical drugs. Malarial parasites are resistant to chloro-

quine after only ten years, and are infecting an ever-widening population of mosquitoes as global warming expands their habitat.

Animals in our care are regularly and routinely dosed with toxic
deworming medication to prevent a range of debilitating symptoms. High-caliber competition horses are dewormed daily to avoid
any depletion in their stamina. Sheep can expect seven different deworming treatments in a year, in changing combinations and rota-
tions, as we attempt to halt the inevitable chemical resistance of parasites. Interestingly, although free-ranging wild animals usually carry
some parasites, they rarely show ensuing symptoms. Chimpanzees at
Gombe usually carry low numbers of one to six worm species and a
few malarial parasites, but rarely show signs of heavy parasite load.
Gorillas too remain outwardly healthy while about half of them
carry low doses of roundworms or hookworms. And when a random
assessment of pancake tortoises in Tanzania was carried out, veterinarians found no blood parasites and only a few nematode eggs in
their feces. Similarly, in the temperate climes of Europe, wild badgers
have nematodes and tapeworms without showing signs of ill health.
In the more polar regions of Argentina, no internal parasites could
be found in rock-hopper penguins. Even in the seas, free-ranging
wild fish carry few parasites while their commercially farmed cousins
are prone to excessive infestations.[1]

Given the ubiquitous and damaging nature of parasites, it is interesting that wild animals manage to do so well. In the past, this observation has led to a belief that internal parasites make no attempt to
push their luck; that they and their hosts do a minimum amount of
harm to each other, in a kind of conspiratorial alliance. But nothing
could be further from the truth. As hosts evolve ways and means of
dealing with parasites, parasites evolve more efficient means of getting around those defenses. When an animal is weakened in some
way by drought, famine, or social stress, the numbers of internal parasites flare up. Any letup in defense will allow a rapid and dramatic
increase in parasites.[2]

Most research into parasite resistance has focused on the immune
system's capability. But the way an animal *behaves* is the first line of
defense against parasites, protecting the immune system from being
overburdened. One reason the role of animal behavior in combating
parasites has not been explored until recently is that parasite-host in-

teractions have usually been studied in laboratories, where the immune response is the *only* response a scientist sees. By carefully observing the *behavioral* strategies animals have evolved, we may learn ways of minimizing the current crisis of parasite control.

Not surprisingly, animals have found ways to avoid parasite hot spots. Since the eggs, cysts, or larvae of internal parasites often pass out of the body with the feces, avoiding feces is one way of avoiding parasites. The Gombe chimpanzees, as we have seen, seem to have "an almost instinctive horror of being soiled with excrement" and only very rarely touch feces (their own or another's) with their bare hands.[3] Rabbits, horses, sheep, and cattle all avoid grazing near droppings, or on grass sprayed with manure fertilizer. Horses at pasture deposit dung and urine in dedicated areas, well away from the grass they graze. Rabbits create special latrines made of pile upon pile of droppings. These act as scent-marking information posts, as well as hygienically separating food from feces. Those species that habitually bring food back to their lairs, such as dogs and cats, go to great lengths to avoid defecating on their own doorsteps. This explains why their domestic relatives are relatively easy to housetrain — they have a natural disposition to avoid soiling their living space with potential parasites.[4] One of the most unnatural things about captivity for many species is the inability to isolate their toilet arrangements.

Yellow baboons in Amboseli National Park sleep on a branch or rock outcrop a few meters off the ground. As they defecate, parasite eggs drop to the ground and hatch two to eight days later. The baboons, which have few favored sleeping groves, move on after one or two nights to a clean site, returning to the original grove only after dung beetles have consumed the fecal matter and thereby reduced the chances of reinfection. Mangabeys, too, change their feeding locations according to the buildup of fecal contamination. Camel pastoralists in Kenya say their nomadic lives arose as a method of keeping their animals free of parasites — an observation borne out by the fact that domestication and intensive housing of the camel's close relative the alpaca creates numerous problems with internal parasites. This natural strategy of constant moving has been adopted in the organic husbandry of farm animals, by rotating stock around different pastures. The need to avoid parasites is thought to play a role in phenomena such as territoriality, migration, and mistrust of strangers.[5]

## SELF-MEDICATION —
## CIRCUMSTANTIAL EVIDENCE?

Nature's pharmacy provides numerous anthelmintic plants capable of harming parasites directly. To be effective against internal parasites they have to be toxic — often highly toxic — so it is curious that many are regularly eaten by wild animals. In 1978 Dan Janzen became one of the first scientists to propose that animals might be using toxic plant compounds to control internal parasites, citing anecdotal observations in which animals such as civets, colobus monkeys, elephants, bison, pigs, tigers, bears, wild dogs, rhinoceroses, Indian mole rats, and jackals seemed to eat plants for their medicinal properties — in particular for their activity against internal parasites. He cited Indian wild bison, which have a habit of feeding on the bark of *Horrahena antidysentaria,* used by local people to treat dysentery. This bark contains an alkaloid active against the endemic amebic dysentery protozoa.[6]

Janzen described how the Asiatic two-horned rhinoceros eats so much of the tannin-rich bark of the mangrove *Ceriops candoleana* that its urine turns dark orange, and he pointed out that the common antidysentery formula Enterovioform comprises about 50 percent tannin. High concentrations of tannins are usually deterrent to mammals (their astringency being disruptive to digestion), but this astringency can harm intestinal parasites more commonly known as worms. The reason tannins are astringent is that they bind proteins, and worms are made of protein.

Support for this idea comes from more recent research. If domesticated goats are fed polyethylene glycol that deactivates tannins in their diet, their numbers of intestinal parasites increase. Furthermore, given a choice, deer do not select food with the lowest possible tannin levels, but instead choose those containing moderate amounts — suggesting that a certain amount of tannin is attractive to them. And when commercially raised deer in New Zealand were fed on tannin-rich plants such as chicory, farmers found they needed to administer less chemical dewormer.[7]

Among Janzen's other examples were elephants, baboons, and silverback jackals feeding on the fruit of *Balanites aegyptiaca*, which local people use as an effective treatment against worms. Since then, scientists have studied two populations of baboons in Ethiopia, along the river that traverses the Awash Falls — one above the falls, one below. The tree is found throughout the range of both groups, yet those baboons below the falls eat the fruit whereas those above do not. Only the baboons below the falls are exposed to schistosomiasis, a blood protozoal parasite spread by water snails. People in this area use the fruit to kill schistosome-carrying snails, suggesting that the baboons too may eat *Balanites* fruit to reduce the impact of schistosomiasis.[8]

Semidomesticated camels in Kenya browse on a popular medicinal plant, *Albizzia anthelmintica*, and in laboratory tests extract from this bark kills 85–100 percent of tapeworms in rats.[9] Camels may also benefit from browsing the leaves and twigs of *Salvadora persica*, a plant that is high in salt and known for its strong purgative effect. Camel herdsmen regularly take their herds to eat these salty plants so that the camels can purge themselves of worms, and they treat themselves in a similar way by drinking from the salt-rich wells.[10]

Purging is one way in which meat eaters can rid themselves of gut parasites. Those fruits that are purgative act by increasing the speed of gut motility. Tigers only occasionally eat fruits, and when they do it is often a purgative fruit. In India they eat the sweet date-like fruits of *Ziziphus jujuba* and the golden berries of Carey's myrtle bloom, *Careya arborea*. The *Ziziphus* contains purgative quinones and *Careya arborea* contains betulin and betulinic acid, active against a variety of parasites and viruses. Jackals also eat *Ziziphus jujuba* fruits, and wild dogs (*Cuon rutilani*) eat the fruits of *Careya arborea*.[11]

Fecal analysis of mantled howler monkeys living in the rain forests of Costa Rica has unearthed another possible example of dietary control of internal parasites. The monkeys are infested with different quantities of internal parasites, depending on where they live. Monkeys in an area called La Pacifica have high levels of infestations, while those living in a different part of the forest, Santa Rosa, have low levels. The most obvious difference between the two sites is the availability of fig trees (*Ficus* sp.). The heavily infested group has no access to fig trees, while the less infested group has many fig trees available. South Americans traditionally use fresh fig sap to cure themselves of worms, as the sap decomposes worm proteins.[12]

There are many other reasons why monkeys might eat figs. They are a highly prized calcium-rich food for many species, including most primates. But we are left with the intriguing idea that the monkeys' diet of fig leaves and fruits may be contributing to parasite control.

When primatologist Karen Strier first started to study the endangered muriquis (or woolly spider monkeys) in Fazenda Montes Claros Park in southeastern Brazil, she was amazed to find them free of all intestinal parasites. This discovery was so startling and so unexpected that she repeated the sampling over many seasons, while checking whether brown howler monkeys living in the same area were similarly parasite free. She found both species completely free of intestinal parasites in this area, but at another location in Carlos Botelho State Park, São Paulo, both species were infested with at least three species of intestinal parasites. The main difference between the monkeys in the two locations was that the worm-free monkeys had access to a greater selection of plants used as anthelmintics by local Amazonian peoples.[13]

In Brazil the long-legged, gold and red, maned wolf (known as the fox-on-stilts) roams the forest at night hunting small rodents, reptiles, ground birds, fish, frogs, and insects. Although it is classified as a carnivore, up to 51 percent of its diet is plant material. By far its favorite is the tomato-like fruit of lobeira (or wolf's fruit), *Solanum lycocarpum*. Although this fruit is more plentiful in certain seasons, the wolf works hard to eat a constant amount throughout the year, suggesting that this fruit has some significant value. What is it that the wolf particularly desires — especially when other fruits are more readily available?

As the maned wolf is endangered throughout most of its range, attempts have been made to breed the species in captivity. Early efforts did not go well, and even today successful breeding is limited to a few experienced teams. Often, infants die young and adults are struck down with unusual cancers. In Brazil one of the most serious health problems is infestation with the giant kidney worm (*Dioctophyme renale*) which, truly deserving of its name, reaches 100 centimeters in length and 12 millimeters in width. As it grows, it eventually destroys the kidneys and kills its reluctant host. It is thought that wolves get the worms from an intermediary fish host. The only known treatment is surgical removal, so infection is usually fatal. However, researchers at the Brasília Zoo found that when packs of captive wolves were fed lobeira daily, they survived. If lobeira was withheld, all the wolves died. Postmortem examinations revealed lethal giant kidney worm infestations, and it was suspected that lobeira might be controlling these worms in the surviving wolves.[14]

As many people around the world are burdened with kidney worms, the potential of lobeira for combating this parasite needs to be explored further. Lobeira is a member of the Solanaceae family, related to deadly nightshade and potatoes. Although the precise constituents of lobeira are still being analyzed, Solanaceae plants contain steroidal alkaloids such as saponins and diosgenin that are active against schistosomiasis and against the malaria parasite plasmodia.

Solanaceae plants often cause visual disturbances and hallucinations. In Brazilian folklore, the mere gaze of a maned wolf is said to be enough to kill a chicken, and the left eye of a maned wolf, removed *while the animal is still alive,* is thought to be a good luck charm — though not for the wolf! Parts of the wolf's body are considered not only magical but medicinal, possibly because they are high in bioactive metabolites from the Solanaceae in its diet. Certainly, metabolites of Solanaceae plants are capable of being stored in a mammal's flesh. Rabbits in England are able to eat large quantities of deadly nightshade (*Atropa belladonna*) with impunity, but their flesh, if eaten by humans, is as poisonous as the plant itself.[15]

There are many possible reasons why animals choose to eat what they do, and the fact that an animal eats something that *could* control parasites does not mean necessarily it is self-medicating, or indeed gaining any incidental benefit beyond nutrition. We need to know whether an animal actually has internal parasites at the time of consumption, and if so, whether they are noticeably reduced afterward. Is there any evidence that animals seek remedies specifically to cure their parasite problem? The story of how scientists explored chimpanzee behavior with this question in mind is an excellent case study that illustrates the many difficulties of investigating self-medication in wild animals.

## CHIMPANZEES AND THEIR WORMS

Wild chimpanzees, studied in Africa for more than forty years, have provided many of our most detailed observations of self-medication in the wild. The Gombe chimps are hosts to numerous parasites, the main internal ones being nematodes (threadworms, whip worms, and nodule worms), one genus of trematode, and numerous protozoa. In the Mahale Mountains, south of Gombe, it is the nodule worms that present the biggest health hazard to chimpanzees, because even a moderate infection can cause diarrhea, malaise, weight loss, anemia, abdominal pain, and lethargy. More severe infestations cause hemorrhagic cysts, septicemia, and blocked colon, clearly visible as large abdominal lumps. Yet such severe symptoms are rare, because chimpanzees have evolved ways to deal with these unwelcome guests.

Early one morning in July 1972, anthropologist Richard Wrangham, then a research assistant to Jane Goodall at Gombe, was following two adult males, Figan and Hugo.

> Walking silently and slowly on the dry paths, Hugo surprised me after three minutes by taking a detour to his right, passing through a thicket and forcing me to crouch and push through the thorny vegetation. Figan followed also, still chewing his palm nuts, and continuing to do so for 10 minutes after he left his tree. At 07.11h, Hugo stopped, and started picking leaves of *Aspilia rudis*. It was odd. He "ate" them very slowly. At 7.20 he left the *Aspilia* patch, whose location he had evidently known in advance, and returned downslope to the eastward path.[16]

The *Aspilia* leaves were selected carefully, unlike normal feeding when bunches of leaves are greedily stuffed into the mouth. Furthermore, they seemed to be kept in the mouth for some time before being swallowed. By the end of that month, Wrangham had seen this strange feeding behavior three times — always soon after dawn. It was obvious that the leaves were not palatable, because often the

A chimpanzee in the Mahale Mountains selects an *Aspilia* leaf to fold and swallow as a scour for worms.
*Michael A. Huffman*

chimpanzees wrinkled their noses as they swallowed them. When Wrangham tried some himself, he found out why. They were rough, sharp, and "extremely nasty to eat."

The primatologist Toshisada Nishida saw similar behavior at Mahale. Although *Aspilia* did not form part of the normal diet, chimpanzees would go out of their way to find *Aspilia* leaves. They would lick, taste, and hold a young leaf on the tongue for a time, often while still attached to the plant, then perhaps abandon that leaf and try another one. When a leaf was finally chosen, it was folded concertina fashion and held in the mouth for a few seconds before being swallowed whole. Later, the undigested leaves reemerged in the feces. Interestingly, scientists noted that feces contained undigested leaves far more often in the rainy season (November to March) than in the dry season; and females swallowed significantly more leaves than males did.

Wrangham and Nishida were intrigued to discover that *Aspilia* plants (Asteraceae) are commonly used in traditional African medicine for treating stomachache and cough. Some kind of self-medication certainly seemed likely, and when *Aspilia mossambicensis* leaves collected from Mahale were analyzed by phytochemist Eloy Rodriguez, the results showed that they contained the chemical thiarubrine-A. Thiarubrines had been discovered recently in other plants and were known to be antibacterial, antifungal, and anthelminthic.[17] They constituted potentially very strong medicine, but could enough active ingredient be consumed by swallowing the leaves? Thiarubrine-A breaks down in the presence of light and in acidic conditions, so it was hard to imagine how it might work inside the body. There was still no evidence that sick chimpanzees used *Aspilia* medicinally — or that if they did, they got better after doing so.

Then in 1987, Japanese scientists at Mahale saw chimpanzees swallow leaves from another species of plant, *Lippia plicata*. Like *Aspilia*, this plant is used medicinally by local people in the area — this time for stomach upsets and menstrual cramps. It has a rough texture like *Aspilia*. What was it about these leaves that the chimpanzees were after? Of all the complex medicinal compounds present, which were the key players? The chimpanzees probably folded the rough-textured leaves to make swallowing easier. Was the texture important? If so, why did chimpanzees not swallow the leaves of *Ficus exasperata*, known locally as African sandpaper — the roughest, hairiest leaves around?[18]

Within the year, Nishida saw chimpanzees swallowing leaves of African sandpaper, as well as several new species such as *Commelina*. Although all the leaves swallowed were known to be medicinal, they were very different chemically from *Aspilia*, as well as from each other. There seemed to be no common denominator for any possible medicinal effects of leaf swallowing. Although the evidence that chimpanzees were swallowing these leaves for medicinal purposes was getting stronger, exactly how, or why, was getting *less* clear day by day. Why did the chimpanzees not chew the leaves? Why fold them and swallow them whole? As the leaves came out the other end *undigested* with only a few surface cells damaged, how could the chemicals inside the leaves have gotten out? It was suggested that the chimpanzees might be buccal rinsing (as humans get a quick fix of cocaine by rinsing coca leaves around the mouth). In this way leaf compounds could be absorbed directly into the blood stream via the mouth, protecting delicate compounds such as thiarubrine-A from harmful stomach acidity.[19] However, Wrangham and Rodriguez were confident that the leaf structure of *Aspilia* was strong enough to protect the thiarubrine molecules as the leaves passed through the acidic stomach, and that thiarubrine-A would therefore be able to act medicinally in the intestine. They also calculated that, theoretically, the amount of thiarubrine getting through would be sufficient to kill 80 percent of a chimpanzee's nematodes.[20]

Nodule worms present their greatest threat at the beginning of the rainy season, which is when chimpanzees most frequently swallow leaves. The seasonal correlation suggested that something in the leaves — possibly thiarubrines — was being used to deal with nematodes. However, Jon Page and Neil Towers meticulously replicated the chemical analyses of *Aspilia mossambicensis* and could find only small traces of thiarubrine in the roots, and none in the leaves. The evidence for a chemical basis to leaf swallowing was receding.[21]

By 1993, chimpanzees across Africa had been seen swallowing the leaves of nineteen different species of plants from widely different plant groups containing an array of secondary compounds with varying chemical actions, many of which had no effect whatsoever on internal parasites. It became increasingly evident that the *only* thing these leaves had in common was their rough texture.[22]

When Michael Huffman looked at freshly excreted swallowed leaves, he found something very interesting indeed. Some of the nodule worms, alive and wriggling, were attached to tiny barbs all over the leaf surface. They were not being killed by chemicals, but were captured by the roughness of the leaves. He was seeing the first example of mechanical expulsion of worms by ingestion of a plant — later to become known as the velcro effect. The reason for the careful concertina folding of the leaves now became obvious: it would increase the chance of hooking worms as they wriggled and became trapped in the folds.

Leaf swallowing has now been seen in at least eleven different populations of chimpanzees, as well as in bonobos and eastern lowland gorillas, in at least ten different sites across Africa. Great apes swallow a variety of leaves from thirty-four species of herbs, trees, vines, and shrubs; some of the leaves have bioactive chemicals, others do not. But all are rough in surface texture, with hook-like microstructures called trichomes. Finding leaves that have these characteristics is no easy task in a forest where most leaves are smooth.[23]

Many of the apes seen swallowing leaves are obviously suffering from symptoms of nodule worm infestation: diarrhea, malaise, and abdominal pain. In each bout, apes swallow from one to a hundred leaves. Leaf swallowing is thought to be particularly effective against nodular worms because they move around freely in the large intestine looking for food and mates, and have no permanent attachment to the intestinal wall. Other worms (such as thread worms and whip worms) burrow into the mucosa of the small intestine and would probably escape the scraping effect of rough leaves. However, leaf swallowing has also helped chimpanzees at Kibale National Park in Uganda to rid themselves of a particularly heavy outbreak of tapeworms (*Bertiella studeri*).[24]

In addition to hooking loose worms, the rough leaves swallowed on an empty stomach stimulate diarrhea and speed up gut motility, further helping to shed worms and their toxins from the body. Furthermore, Huffman deduces that when adult worms are removed from the gut, larvae emerge from the tissues, thereby relieving general discomfort.[25] This fact is significant in that it takes us away from the idea that chimpanzees have found a way to target parasites, toward the more familiar ground that they are seeking relief from uncomfortable sensations.

## OTHER ANIMALS SCOUR THEIR WORMS

Once physical scouring became the accepted hypothesis to explain leaf swallowing, it became clear that primates are not the only species to use this method of worm control.

At certain times in an animal's life, such as migration and hibernation, parasites are more of a health hazard than usual. If an animal goes into hibernation with a gut full of parasites, the parasites will eat the animal's carefully stored food reserves while it sleeps. Biologists have known for years that hibernating bears somehow get rid of internal parasites in the autumn before hibernation. When Barrie Gilbert spent six years studying Alaskan brown bears in Katmai National Park, he noticed that they changed their diet and started to eat "strange things" in the few months leading up to hibernation. Highly fibrous sharp-edged sedge (*Carex* spp., Cyperaceae) began to appear in large dung masses otherwise almost completely composed of long tapeworms. The coarse plant material was apparently scraping out the worms in a way similar to the rough leaves swallowed by chimpanzees. Physical expulsion seems to be used by Canadian snow geese as well. The geese carry significant tapeworm burdens in the summer, but just before migration they deposit large boluses of undigested grass and tapeworms in their dung. When they reach their migration destination, they are completely clean of tapeworms. In both brown bears and snow geese, worms are shed at a time of critical nutritional stress, a time when carrying these parasites would greatly reduce the animal's chances of survival.[26]

Closer to home, domestic dogs and cats both occasionally chew grass, as every pet owner knows. The grass seems to have two effects: one is emetic (stimulating regurgitation or vomiting), the other, a purgative scour (ridding the body of worms farther down the intestine). Thus the grass could work at either end of the intestine, depending on which orifice was nearest to the problem. Herbalist Maurice Mességué claims that some dog species discriminate between different grasses for different medicinal functions, using hairy grasses for emetics and couch grass as a purgative. In Tanzania a veterinarian known to Huffman observed a dog shed roundworms by eating grass.[27]

Our companion cats and dogs are most likely carrying on a residual self-medication strategy of their wild ancestors. Wolves eat grass — and all year round, it seems. Biologist Adolf Murie, studying a pack on Mount McKinley, noted that grass eating seemed to act as a scour, for roundworms often came out with the grassy droppings. He watched while one particular wolf ate grass for a few minutes and then produced a watery scat. Later the same wolf vomited some of the grass he had eaten. According to Indian folklore, tigers will (on rare occasions) eat grass "when hungry" — and if heavily infested with worms, the tigers may appear emaciated. A small portion of the droppings of wild Indian tigers has been found to consist almost entirely of grass blades, and in at least one case a tapeworm was found inside.[28]

Traditional herbalists have used purges and scours for thousands of years as a method of worm control, finding it safer than toxic anthelmintic preparations. A delicate balance exists with chemical dewormers, between a dose toxic enough to kill or shed parasites, yet not so toxic that it harms the host. These kinds of physical remedies may be a particularly useful addition to parasite control in modern farming, where parasites are becoming increasingly resistant to chemical treatments.

## MORE ON THE CHIMPANZEES

On November 21, 1987, at the beginning of the rainy season, Michael Huffman and Mohamedi Seifu Kalunde were in the Mahale Mountains watching chimpanzees feed.

The chimpanzees sat quietly plucking the red wax-coated seeds of Lulumasia, a relative of nutmeg, from the yellow husks clustered like grapes. The two men located the chimpanzees in the dense vegetation by listening for the sound of discarded husks falling to the ground with a light thud. To the accompaniment of pant-grunt greetings and screams, they came across a small group of three adult females and their four youngsters. One of the females, Chausiki, was clearly unwell, and she slept while the others fed. When awake, she moved slowly and reluctantly, ignoring the pleadings of her young son, which normally would have had her rushing to his aid. Her urine was dark and discolored, her stools loose, and her back quite obviously stiff. The two men followed Chausiki to a small shrub, *Vernonia amygdalina*. This plant, commonly known as bitter-leaf, is so poisonous that it is called "goat killer" by the Temme people of Sierra Leone. The extreme bitterness successfully warns most animals to stay away. But not this sick chimpanzee. Chausiki bent down several shoots of bitter-leaf and carefully stripped off the outer layers to reveal the inner pith, which she chewed and sucked for at least twenty minutes, making audible slurping noises and spitting out unwanted fibers.

This chimpanzee sucks the bitter pith of *Vernonia amygdalina*. Known as "goat-killer," this toxic plant has strong antiparasitic properties.
*Michael A. Huffman*

Fortunately for Huffman, his companion not only had thirty-five years' experience tracking chimpanzees, but was also a Tongwe traditional herbalist. As they watched Chausiki chewing the bitter pith, Seifu explained that the plant was a very strong dawa (medicine) for local people, used to treat malarial fever, stomachaches, schistosomiasis, amebic dysentery, and other intestinal parasites. Furthermore, pig farmers in Uganda supply their animals with branches of this plant, in limited amounts, to treat intestinal parasites.

Chausiki continued to suck on the bitter-leaf pith while other healthy chimpanzees ate far more nutritious and palatable plants. Her son Chopin begged, as usual, for some of what she was eating, but she ignored him. Eventually he got hold of a piece she had dropped and eagerly put it to his mouth. He quickly spat it out in obvious disgust. For the remainder of the day Chausiki took frequent long naps, eventually making a night nest unusually early. Next morning, she was still visibly weak, frequently stopping to rest or sit still, but after a long midday nap, she appeared to be on the mend. Traveling swiftly through the dense forest, she left the rest of the group far behind. She even regained her appetite, and the men left her that evening feeding on elephant grass.[29]

Later analysis of plants collected at Mahale found that the bitter pith of *Vernonia amygdalina* contained seven previously unknown steroid glucosides, as well as four known sesquiterpene lactones, capable of killing parasites that cause schistosomiasis, malaria, and leishmaniasis — any one of which could cause the symptoms seen in Chausiki. The sesquiterpene lactones (previously known to chemists as "bitter principles") are not only anthelmintic but antiamebic, antitumor, and antimicrobial. The outer bark and leaves, which Chausiki so carefully discarded, contained such high levels of vernonioside B1 that they would have been extremely toxic to a chimpanzee. It seems that not only had she picked a suitable plant to deal with her symptoms, but she had found the right *part* of the plant to be effective without harm to her.[30]

Bitter-pith chewing appears to be rare, but by 1997 four other chimpanzees with diarrhea, malaise, and nematode infection had been seen chewing it. Two of these individuals recovered within twenty-four hours (similar to the recovery time of local Tongwe people using this medicine). Without doubt, the behavior impacted on the nodule worm infestation. In one case, the fecal egg count dropped from 130 to 15 nodular worm eggs within twenty hours of chewing bitter-pith. In each instance, chimpanzees took a detour from their normal feeding forays to specifically find bitter-pith plants. Bitter-pith chewing, like leaf swallowing, is more common at the start of the rainy season, when nodular worms increase.[31] Furthermore, Huffman has noticed that chimpanzees with higher worm loads, or those that appear to be more ill, tend to chew *more* bitter-pith than those with lower infestation levels.

Bitter-pith chewing fits all the criteria for curative self-medication: a sick animal seeks a rarely used plant with little or no nutritional value, prepares it or feeds on it in an unusual manner, continues the behavior while it is sick but stops when it is well, and recovers within a plausible time. In addition, the plant contains a compound capable of having a medicinal effect, and a potentially effective dose is consumed.

Curative measures such as this are likely to be rare because they are used only as a last resort during periods of intense discomfort. The abdominal pain, diarrhea, and bowel irritation of nodular worm and tapeworm infestations could be the triggers for leaf swallowing or the chewing of bitter-pith. Meanwhile, the everyday diet contributes much to the sustainable control of parasites. Chimpanzees at Mahale, for example, eat at least twenty-six plant species that are prescribed in traditional medicine for the treatment of internal parasites or the gastrointestinal upset they cause.[32]

Chimpanzees could be using simple cues such as hairiness or bitterness to find substances that make them feel better quickly. Such positive postingestive feedback is at the root of many aspects of diet selection, which can vary from moment to moment with changing internal conditions. It certainly seems that bitterness in plants may be an effective indicator of medicinal properties: it generally indicates toxicity, but it is this very toxicity that is so useful against parasites. *Vernonia amygdalina* is not just bitter, it is the most bitter plant the chimpanzees can find in the forest — and the *only* very bitter plant chimpanzees have been seen to ingest. One slurp of its juice will make an adult human wince. Chimpanzees and other animals normally avoid it, but appetitive or tolerance changes may take place during sickness. Jane Goodall discovered this at Gombe, when she wanted to help some sick chimpanzees by lacing bananas with antibiotics. She was worried that if chimpanzees that were not ill got the medicine, it would disrupt their essential gut microflora; but as it transpired, only the sick chimpanzees ate the bitter-tasting antibiotic bananas. The healthy chimpanzees avoided them and waited for her to put out untainted, sweet bananas.

Traditional herbalists have for generations used a patient's perception of bitterness as an indication of ill health. Sick human patients are tolerant of extremely bitter herbal remedies, but as they get better their tolerance of bitterness declines and the prescription has to be altered accordingly.[33] The mechanism that brings about these changes is not yet known, but experimental evidence supports the idea of a bitter-seeking "rule-of-thumb" behavior. Mice infected with malarial parasites were provided with a choice between two water bottles; one contained only water and the other a bitter-tasting chloroquine solution that would combat malarial infection. Control mice were given only water. Infected mice with access to chloroquine experienced significantly less infection and mortality than control mice. Malarial infection was reduced because the mice took approximately 20 percent of their fluid intake from the water bottle containing the bitter chloroquine solution. Consumption of chloroquine was *not* related to the malaria infection, however. Given a choice, both sick and nonsick mice took small doses of the bitter solution. Many other apparently healthy mammals also sample a diverse range of bitter substances, a habit that may represent a generalized behavioral strategy of preventive self-medication against parasitic infections and other illnesses. Whether the baseline bitter appetite changes with different ills has yet to be explored.[34] Pain (as we saw in Chapter 7) certainly appears to enhance bitter tolerance in rats and chickens seeking analgesia.

Chimpanzees do not need to understand that they are infected with worms in order to medicate themselves effectively. They simply need to feel gastrointestinal malaise, and link relief to bitter-pith chewing or leaf swallowing. They are more than capable of learning such a simple association. Yet some chimpanzees do seem to be aware of having worms. When one male, followed by Huffman, was attempting to rid himself of an overwhelming nematode infestation by chewing bitter-pith and swallowing leaves, he actually caught and held some of the worms as they fell from his bladder. Not wanting to pass up a source of protein, he ate them — showing perhaps that he did not realize the worms were the cause of his malaise.[35]

Because the parasites that chimpanzees have to deal with also infect people, pigs, sheep, and cattle, a greater understanding of the way they deal with them could have many advantages. Local home-grown strategies could free farmers in developing countries from the grip of international pharmaceutical companies and their expensive factory-made drugs. Research is continuing on the efficacy of *Vernonia amygdalina* and a number of the other plants in the chimpanzee diet for possible control of schistosomiasis, leishmaniasis, dysentery, and drug-resistant malaria.[36]

As studies of primate behavior continue, no doubt new strategies for self-medicating will be revealed to us by these intelligent creatures. We have so far had only a glimpse into the world of primate health maintenance, throwing a glimmer of light on the origins of human medicine 10 million years ago, when our ancestors split from our nonhominid relatives.

## EVEN INSECTS DO IT

It is not only primates, or even mammals, that use plant medicines to control parasites. It has long been known that certain butterflies harvest and store the toxic cardiac glycosides from milkweed plants, and that this stash protects them against some predatory birds. However, in 1978 it was discovered that the cardiac glycosides also protect butterfly larvae from internal parasites.[37] It is not clear whether these effects are merely incidental to feeding; however, the dietary choice is distinctly beneficial.

Scientists studying insect parasitoids (lethal parasites) have found convincing evidence that insects do self-medicate. Woolly bear caterpillars of the tiger moth spend the spring munching vegetation. By summer, when parasitic tachinid flies inject their eggs into unlucky caterpillars, they are plump and succulent. The fly larvae develop inside the caterpillars' abdomens, feeding off their fat reserves and eventually taking up the whole body cavity. Finally, the larvae emerge by making a hole in the cuticle wall. When studied under laboratory conditions, most caterpillars (quite understandably) die from this experience; but when Richard Karban and his colleagues started rearing their caterpillars in outdoor field cages, they noticed that the survival rate of parasitized caterpillars was much higher.[38]

Outside, the caterpillars ate two main food plants, lupine (*Lupinus arboreus*) and hemlock (*Conium maculatum*). While healthy caterpillars fared best when they were fed on lupine, parasitized caterpillars did better on the more poisonous hemlock. Given a *choice* of plants, healthy caterpillars preferred to feed on lupine and parasitized caterpillars preferred to feed on hemlock. In other words, having parasites affected diet choice, *and the change in diet improved their chances of survival.* Although hemlock (known to contain at least eight alkaloids) does not kill the parasites, it helps the caterpillars survive the infection. We do not have to imagine a complex cognitive mechanism for insect self-medication — merely that the action rapidly reduces a survival threat. Once again we see the importance of studying animals *under natural conditions*. If Karban's team had continued to do all their research in the laboratory, the caterpillars would never have had the opportunity to display their self-medicating strategies.

Bumblebees are also known to change the flowers on which they feed when infected with parasitoids, and trematode infections alter dietary preferences among freshwater snails — although in neither case has it been established that the host benefits. We know that some of the most voracious and damaging pests of human food crops are insects, so understanding how they deal with their own parasites could help us to sabotage those strategies and thereby protect our food crops.

Temperature control is another way of dealing with parasites. One successful strategy by which bumblebees relieve the symptoms of their lethal parasites (conopid flies) is staying cool. Parasitized worker bees (*Bombus terrestris*) remain in the field overnight rather than returning to the nest with the others, and the cooler temperature retards the development of the parasite.

Other invertebrates employ similar strategies. Freshwater snails infected with the trematode *Schistosoma mansoni* seek out pockets of water with lower temperatures. On the other hand, grasshoppers infected with protozoan parasites, and fish infected with immature tapeworm eggs, seek warmer temperatures than their uninfected fellows, thereby increasing their chances of survival with an immune-enhancing fever.[39]

## ANOTHER ROLE FOR EARTH

In parasite control, as in other health-maintenance strategies, earth eating plays a role. Lambs and other animals infested with worms often consume earth.[40] Clay is especially useful. It can help in three ways: by adsorbing toxins secreted by the parasites, by physically expelling worm eggs, and by protecting the gut from invasion by migrating worm larvae. In tropical Africa, people infested with hookworm eat clay in an effort to relieve their symptoms of gastric irritation.[41]

An actively managed, high-density, free-ranging group of rhesus macaques on Cayo Santiago Island, Puerto Rico, are quite heavily infested with internal parasites, particularly the nematode *Strongyloides fuelleborni*. (This population density is artificially high because of feeding.) In captivity, rhesus macaques have to be regularly dewormed with strong chemical anthelmintics. The usual symptoms of these worm infestations are diarrhea, dysentery, debility, and even death. Although 89 percent of the Cayo Santiago monkeys are infected, only 2 percent have diarrhea, and *all appear healthy*. Reproductive rates are high and mortality rates low.

These parasites do not swim around in the intestine like nodule worms in chimpanzees, so the physical scraping of leaf swallowing would not be effective. Instead, the macaques consume clay on a daily basis, and in their quest for it they have excavated dozens of mine sites scattered throughout the island. Old mine sites are abandoned and new ones created as the clay they seek, as much as 30 centimeters below the surface, is used up. The local clay is a highly adsorbent form of kaolinite and smectite that counteracts the fluid loss and subsequent debility of chronic diarrhea. Kaolinite adsorbs toxins and bacteria and forms a protective coating on the inside of the intestine, which may be particularly useful for dealing with parasites such as these that cause ulceration and bleeding from the gut.

Female macaques eat more earth than males, and they have fewer parasites than males. Older monkeys also have fewer parasites than younger ones; again, this is thought to be a result of the earth they have eaten. If soil is consumed over a long period, the gut wall starts to thicken. This is one of the problems of long-term geophagy in humans; but in this situation, when worms burrow through the gut wall, the thickening could be beneficial. Eating clay appears to be both a curative remedy for the symptoms of parasite infection and a preventive against recurring infection.[42]

We have seen only a few of the ways in which animals keep parasite infestations under control, even though they do not seem to achieve all-out eradication. Most endure a persistent low level of parasite infection — which may incidentally stimulate the immune system to stay on guard. It may be adaptive to tolerate a residual number of internal parasites rather than to expend resources and energy to eradicate them, especially in an environment where constant reinfection is probable. However, in times of extra demand (such as pregnancy, lactation, migration, and hibernation) it may be worth making additional efforts to expel the unwelcome guests.

Research suggests that regular exposure to low doses of intestinal worms may even be protective against modern health problems such as inflammatory bowel disease. One theory is that in developed countries many of us live in worm-free environments, and our immune systems (which evolved to do constant battle with unwelcome guests) become overly active in the absence of any enemy.[43]

Internal parasites remain a major curse on people and their livestock. The discovery of any successful treatment is therefore likely to be highly profitable for pharmaceutical companies. The steroid glucosides in bitter-pith, revealed to us by chimpanzees, are the first new drugs found by watching animals. However, it is unlikely that animal self-medication will provide many ingredients that can be patented, mass-produced, and marketed: animals are clearly using a combination of avoidance, prevention, and cure to deal with parasites and their symptoms. The lesson to be learned from wild health is that successful parasite control stems from constant vigilance rather than from magic bullets.

# 10

# GETTING HIGH

> The discovery of most of our major psychoactive drugs came from early man's observations of animals.
>
> — Ronald K. Siegel, 1989

IN SOUTH AFRICA, wild baboons go out of their way to find and eat the red plum-like fruit of a rare and poisonous cycad tree, even though other food may be available nearby. They become intoxicated and stagger about, unable to move quickly, quite obviously unaware of any danger. In South America, spider monkeys eat fermented fruits, become boisterous, chuckle, and scream before slumping into a stupor. In North America, small waxwing birds fall to their deaths from rooftop perches, inebriated after feeding on fermented hawthorn berries (*Crataegus*).[1]

Wild animals occasionally indulge in recreational drugs.[2] They get drunk, have hallucinations, sedate themselves into a stupor, and eagerly consume stimulants. What are they doing? Are they accidentally getting high as they forage for food, or could there be physical or psychological benefits to such indulgences?

Nature's pharmacy provides numerous intoxicants: stimulants, sedatives, hallucinogens, analgesics, euphorics, and inebriants. Although some of these substances are found in fungi, bacteria, insects, and other animals, by far the most come from plants. Many have powerful effects on the vertebrate nervous system — usually by mimicking the action of natural neurological chemicals. Plants most likely make these compounds as a defense against their herbivorous predators. Indeed, when laboratory animals are presented with strong intoxicants, most will not return a second time. Still, some animals (both wild and domesticated) not only consume intoxicating plants but do so readily and repeatedly.

In a society blighted by abuse of and addiction to drugs (both prescribed and nonprescribed), we need to know a lot more about why animals interact with natural intoxicants in this way. Although a wealth of laboratory research has been performed on intoxication in domestic species, observing free-ranging animals can teach us much more about how and why the consumption of intoxicants evolved.

## BOOZY BEASTS

When fruits, grains, saps, or honey ferment, the sugars in them turn to ethanol, the type of alcohol found in alcoholic beverages. In certain conditions, mushy fruit cocktails can contain up to 12 percent alcohol, equivalent to wine or very strong beer. All of the animal species that have been studied (from insects to elephants) show the same somewhat familiar reaction to alcohol: initial excitement, followed by an uncoordinated phase, followed finally by sedation. Fruit flies, bumblebees, wasps, hornets, and aphids feeding on fermenting sap or fruit become uncoordinated and temporarily grounded. Sapsuckers (a kind of woodpecker) get wobbly on fermented sap. Indian sloths get pleasantly inebriated chewing on fermented flowers. Migrating birds, flying south for the winter, feed on fermented berries and get so drunk they fall out of the sky — a nightmare for car drivers in Virginia.

Robins and cedar waxwings are by far the most common avian drunkards in North America. In some cases, their health is threatened by their inebriated antics as they crash into windows and overhead wires. Driven south by the harsh Russian winter, flocks of Bohemian waxwings feed avidly on the berries and fruits still hanging on the trees in southern Scandinavia. Their favorite berries, rowans (*Sorbus aucuparia*), have been around so long that they have started to ferment. When the weather suddenly turns cold, small groups of birds are found dead beneath their roosting trees in the morning.

With no apparent injuries or disease, they appear to have simply fallen off their perches. Postmortem examinations reveal that these birds were severely drunk when they died, and that they have acute alcoholic liver disease.[3]

While a drunken insect may be amusing and an inebriated bird a minor problem, drunken elephants on a binge are a terrifying sight. In 1985 a herd of 150 Asian elephants broke into an illegal still in West Bengal and drank copious quantities of moonshine. Inebriated, they rampaged across the land, killing five people, injuring a dozen, demolishing seven concrete buildings, and trampling twenty village huts.[4]

Elephants have something of a reputation for getting drunk. In 1875 W. H. Drummond wrote of the long trek made by African savannah elephants to find a favorite fruit. He noted that after eating the umganu fruit, they became quite tipsy, staggering about, screaming so loud as to be heard miles off, and having tremendous fights. Throughout Africa, elephants get drunk on the fermented, yellow, plum-sized fruit of the marula tree, as well as on doom palm, mgongo, and palmyra palm fruits. Dian Fossey noted that elephants in Rwanda favored a grapefruit-sized fruit that the locals called mtanga-tanga and that the beasts became besotted after extended intoxicating binges. When elephants get a whiff of ripening fruit, they make a rapid beeline toward it, often from more than 10 kilometers away. By the time they arrive, some of the fruit has inevitably started to ferment. Those elephants arriving later will get even drunker than those arriving first. What is more, fermentation continues inside the elephant's gut, producing even greater amounts of alcohol.[5]

A fruit in Borneo is so large and nutritious that many species fight over it. Tigers are even rumored to kill people carrying them; they take off with the fruits, leaving the dead humans behind. The prized fruits are those of the durian tree, famed for its dreadful smell. Monkeys feeding on fermented durian fruits start to shake their heads vigorously and have difficulty climbing. Despite these effects, monkeys repeatedly return to eat the fruit — along with flying foxes, bats, and of course elephants.[6]

Why consume fermented fruit, when inebriation is obviously threatening to survival? It could be that inebriation is accidental — a side effect of feeding on nutritious fruit — but the smell and taste of alcohol are quite pronounced, so theoretically, at least, it could be avoided. Furthermore, there are indications that some individuals may *like* the taste of alcohol.

On the Caribbean island of Saint Kitts, free-ranging wild (but introduced) vervet monkeys like alcohol so much that the local people entice them with beer as bait. Some monkeys have become bingers, regularly drinking themselves unconscious, while others are steady drinkers. Only 15 percent abstain completely. Judicious cross-breeding experiments suggest that these differences may be genetic, and research continues in order to find the gene or genes that predispose individuals to heavy drinking.[7] Laboratory rodents also reveal a built-in liking for alcohol: C57 strains of wild-type mice prefer to drink a 5 percent alcohol solution rather than water (and they will drink to the point of staggering), whereas mice of a different strain living in exactly the same conditions remain teetotalers.

We might expect natural selection to work against a gene that leaves its carriers vulnerable to predators and accidents. The evidence for a genetic predisposition for liking alcohol, and an ability to search for and find it, suggests that there may be an opposing selective force — some *benefit* to alcohol consumption. What could that be?

The genetic predisposition for liking alcohol has been of great interest to scientists studying human alcoholism, which often runs in families. Yet genes are not the only contributory factors; social and developmental factors take part too. Stress, for example, often increases an animal's readiness to imbibe as a form of self-medication. And the conditions under which a monkey grows up affect its drinking behavior later in life. The more stress a monkey experiences in early life, the more alcohol it will voluntarily drink as an adult.[8]

Ron Siegel investigated whether a wild elephant's fondness for alcohol was a form of self-medication against the stress of poachers and tourists. He offered a small group of African elephants in a Californian game reserve access to unlimited amounts of unflavored 7 percent alcohol, under varying social conditions. They readily consumed moderate quantities of alcohol, which soon had them staggering and swaying. When their home range was reduced by half and they had to compete for food with other species, the amount of alcohol they consumed increased dramatically. Then, when the home range was increased again and the competing species were taken away, alcohol consumption returned to normal (lower) levels — suggesting that stress was a contributing factor.[9]

However, there may be other reasons why alcohol consumption increases when food supply is restricted. The evolutionary biologist Robert Dudley does not believe intoxication is the principal aim of alcohol consumption in animals. Alcohol is a useful source of calories, providing roughly twice the energy content of carbohydrates. Ripe fruits ooze a trail of ethanol vapor that attracts animals, and an evolved sensitivity to that vapor would lead an animal straight to an energy-rich nutritional prize. Furthermore, ethanol stimulates the appetite, so it may have stimulated the fruit eaters to consume more fruit and thereby distribute more seeds of the plant. We humans (along with other fruit eaters) may have evolved a genetic predisposition to *like* alcohol because of its energy.[10] From this perspective, the smell of alcohol indicates a source of energy, and getting drunk is merely a hazardous side effect of feeding on an especially ripe batch of fruit. Siegel's elephants might have drunk more alcohol when confined to a smaller home range in order to gain more calories as competition for food increased. The temporary inebriation was the price they paid for those calories.

This energy-seeking theory could explain the huge popularity of durian fruit, despite its dangerous inebriating effects. Durian is packed with minerals, vitamins, and carbohydrates. If the alcohol-drinking C57 mice are given an option of chocolate or sugar, they select those and reduce their alcohol consumption, suggesting that they drink alcohol for its calories, not for intoxication.[11]

Even a search for calories may not be the end of the story. We have learned that moderate alcohol consumption also has medicinal benefits: heart disease and death rates are lower in humans who consume moderate amounts of alcohol than in those who consume large quantities or abstain completely. This could bring us full circle, for one of the ways in which moderate alcohol consumption is thought to protect against disease is by reducing the impact of stress.

Other benefits of alcohol consumption are more direct. Changes in blood fats and clotting factors make coronary heart disease less likely; and paradoxically, a couple of alcoholic drinks a day may actually help heal damaged livers, by activating a gene for a growth hormone that helps regenerate liver tissue. Beneficial effects such as these, combined with the high energy content of alcohol, may account for the evolution of a liking for moderate levels of alcohol. However, as we know only too well, this taste for alcohol can develop into life-threatening alcoholism.[12]

If so many species are attracted to alcohol, why are there not more animal alcoholics in the wild? Quite simply, they have little opportunity to develop a harmful dependency. In nature, the supply of alcohol is relatively scarce or sporadic, and animals that overindulge will be easy prey for predators. Natural selection has not had much opportunity to select out those prone to alcoholism. Now, however, we humans have created an abundant supply and we are suffering the consequences.

## MAGICAL PLANTS

By carefully observing wild animals, humans have discovered not only medicines but powerful potions with seemingly magical powers, which they have used in ritualistic ceremonies or for making contact with the spirit world. As a plant may provide visions of unknown worlds and new physical or psychological sensations, it is easy to understand why people could think the plant itself contained, or had access to, spirits or other worlds.

When European explorers first discovered villages in the forests of Gabon and northern Congo, they found native people cultivating the iboga shrub (*Tabernanthe iboga*) because of its ability to energize, stimulate, and create visions. The villagers explained how they had discovered the plant by watching wild boars dig up and eat the roots. The boars would then go into wild frenzies, jumping around and seeming to flee from invisible foes. When porcupines and gorillas were seen doing the same, the locals knew they had found a powerful plant. The Bwiti people of the Gabon used iboga in their initiation rites. When boys drank the bitter-tasting root scrapings, they would feel great strength and see visions for several hours, and then pass into a deep sleep that could last up to a week![13] Laboratory analysis of iboga has uncovered many alkaloids — most commonly ibogaine, a stimulant that can act like an amphetamine or as a hallucinogen, depending on the dose.

The ancient Mayans are thought to have stumbled onto a powerful narcotic by watching bees. The taste and chemical composition of honey is dependent on the flowers from which the bees have collected nectar. If toxic or narcotic plants supply the nectar, the toxic honey can cause outbreaks of physical and psychological trauma, even death. When Mayans found honey that gave them powerful hallucinations, they watched where the bees went to get their nectar and in that way found plants with narcotic properties. Honey made from the flowers of the *Turina corymbosa* (xtabentun, in Mayan) is powerfully hallucinogenic. Mayans even today set up tree-trunk bee hives near these plants to obtain a mind-altering honey for their narcotic mead, called balché. This hallucinogenic honey is also used to induce labor in childbirth, for the active ingredients, ergoline alkaloids, stimulate uterine contractions.[14] We do not yet know why bees use these particular plants as nectar sources, but it may be that they gain protection from pathogens (as do other insects mentioned in Chapters 6 and 9).

Two signs of intoxication that are easy to see in the wild are stimulation and sedation. The *Journal of the Bombay Natural History Society* in 1941 reported that solitary Malay elephants often fed extensively on a vine called *Entada schefferi,* which appeared to have a stimulating effect. "The elephants would travel great distances after such a meal." On the other hand, after they had fed on a palm called *Oncosperma horrida,* they seemed drowsy and unwilling to travel. "In fact, I have known occasions when an elephant, having fed well, but possibly not too wisely, on this palm would not travel more than a few hundred yards from where there was a stand of this plant before he would lie down for a nap; only to return again on waking up to have another gorge."[15] Eating this palm is no easy matter: it is covered in long, tough thorns that point downward. The elephant has to push over the whole palm with its head, then stamp on the upper half of the tree to expose the inner pith. Even then, it is so difficult to consume much of the pith that it seems unlikely that the resulting intoxication is accidental.

Coffee, one of the most commonly used stimulants in the modern world, was also reputedly discovered by watching animals. Folklore tells that in about A.D. 600–800 a goatherd named Kaldi was tending his flock by the shores of the Red Sea. He noticed that the goats were behaving strangely, romping around energetically and staying awake when they would normally be sleeping. Even the older goats were jumping and leaping the whole night long. This activity started after they had browsed on the red berries of a small shrub, so he tried some himself. He liked the feeling of exhilaration so much that he shared his discovery with monks in a local monastery. After drinking tea made from the berries, the monks were able to stay awake through long hours of prayer. Convinced they had been strengthened by a heavenly offering, they called the drink *kahveh,* meaning "stimulating and invigorating."[16]

Goats are also credited with the discovery of another popular stimulant — khat, or Abyssinian tea. A legendary Yemenite herder called Awzulkernayien noticed one of his goats running at extraordinary speed after feeding on leaves of the khat tree (*Catha edulis*). When he tried chewing the leaves himself, he felt revived even though he was tired from a day's work. The habit of chewing leaves to banish fatigue spread quickly throughout the country. It is now known that the leaves contain amphetamine-like alkaloids that deter most animals — except goats. So strong is their liking for this stimulant that commercial khat fields must be heavily guarded with powerful electric fences to protect the plants from marauding goats.

With stimulants, as with alcohol, it may be energy that animals are after — or, rather, the *perception* of energy. Stimulants that banish fatigue or discomfort may be perceived as high-energy foods, even if they are not. Sedation, on the other hand, may reward an animal simply by reducing goal-seeking behavior.

About 70 percent of domestic cats love catnip, although the percentage among wild cats is unknown. This perennial herb has downy leaves and gives off a minty-alfalfa odor. When cats come across catnip, they bite at it and roll in it, stare and leap, and shake their heads. Small amounts are not harmful, but in concentrated doses catnip is hallucinogenic. Some cats paw and play with phantom butterflies and pounce on invisible prey; others hiss and show fear, even when no other animal or object is present. For genetic reasons, some cats respond to catnip and some do not. Responders have a gene that enables them to detect the active ingredient nepetalactone, a terpenoid. Nepetalactones mimic a natural courtship pheromone found in male-cat urine, which is thought to stimulate a psuedo-sexual reaction.

As we saw in Chapter 8, catnip is also repellent to insects and skin pests. When Ron Siegel offered catnip to civet cats, they showed only mild curiosity, sniffing, sneezing, and rubbing chins. In contrast, a sensitive young tiger took one sniff and leaped a meter in the air, urinating in the process, and fell flat on his back. He scrambled to his feet and dashed head first into the wall of his cage! Catnip is powerful stuff. In the wild, the big cats of Africa have their own version of catnip — catnip leaf, or lion's ear (*Leonotis nepetifolia*). Joy Adamson often saw her orphaned leopard rub and roll delightedly in this plant, and the same plant was commonly used by the Hottentot for its psychoactive properties.

## TRIP OR TREAT?

It is easy to understand why fruit eaters might like the taste of fermented fruit and the alcohol it contains, and why the sexual smells of catnip are attractive to cats. But why would animals be interested in stronger psychoactive plant compounds such as alkaloids? Unlike fruit, which has evolved to be tasty in order to attract animals to eat it and distribute the seeds, alkaloids are usually bitter and unpalatable compounds that evolved to *dissuade* animals from eating a plant. Mysteriously, though, many animals regularly and repeatedly consume powerful psychoactive plants containing bitter-tasting alkaloids, often displaying obvious psychological or neurological effects.

*Datura innoxia* and *Datura stramonium* (also known as thornapple, devil's apple, and jimsonweed) are highly poisonous members of the Solanacae family. Their strong smell and bitter taste deter most animals. They protect their seeds in thorny casings, and their leaves and seeds with narcotic alkaloids such as hyosyamine. They are toxic and hallucinogenic at high doses. Despite this, or perhaps because of it, baboons repeatedly eat small amounts of both *Datura* species, along with another hallucinogen, *Euphorbia avasmontana*. Even insects indulge in the narcotic effects of *Datura* plants. Hawk moths that pollinate the flowers appear intoxicated after feeding on the nectar, yet they return for more, as if unperturbed or even attracted by the effects. Siegel observed: "The moths are normally great flyers, but after drinking *Datura* nectar, they have difficulty landing on flowers and often miss their target completely, falling into the leaves or onto the ground. They appear to have trouble getting up again and when they resume flight, their movements are erratic and disoriented."[17] In herbal medicine *Datura stramonium* is used as an antispasmodic to relieve the symptoms of asthma and other bronchial complaints. In addition, it is sedating and mildly analgesic.

In the Amazon jungle, the Tukano Indians say that jaguars claw and gnaw at the nauseating bark of the yaje, or ayahuasca, vine — sometimes even chewing the leaves, which is strange behavior for a meat eater. The vine contains many alkaloids, one of which, harmine, causes laboratory dogs and cats to jump and stare at objects that are not there. The Indians believe the yaje vine sends them on flights to another world and use it in their shamanic rituals. One action of the vine is to dilate the pupils, enhancing vision and general perception — an action shamans exploit to help them see in the dark with "jaguar" eyes, and hunting Indians use to enhance their visual perceptions in the darkness of the forest. Siegel suggests that sensory enhancement might even be why the jaguar — the forest's prime nighttime hunter — uses it too.

The yaje vine serves not only to encourage visions and to improve night sight. It is also used by local people to rid the body of internal parasites, and it may be this benefit that encourages the jaguar to eat the vine. Phytochemist Eloy Rodriguez has pointed out that most psychoactive plant alkaloids are also potent antiparasitic drugs, and that these plants may be eaten primarily to control internal parasites, intoxication being merely a side effect. The yaje is strongly emetic and purgative, and users claim to be cured of many physical ills after a course of vomiting and diarrhea from this plant.[18]

Coca, a combination of many South American plants, has been chewed by humans since at least 5000 B.C. Peruvian Indians tell that coca was first discovered by watching how llamas selected leaves when carrying packs a long way from their normal forage. The coca's leaves appeared to have a sustaining effect on the llamas, so their human companions tried them too. Other wild animals, such as sloths and monkeys, eat coca leaves. Although these leaves contain a small amount of the alkaloid cocaine, which alleviates hunger and fatigue, they also contain many nutrients, minerals, and vitamins. They are a fine choice for a llama in the middle of a hard day's work.

Some intoxication is most likely accidental as hungry animals try out a variety of foods. When herds of wild caribou migrate across Canada, a few individuals nibble on clusters of the fly agaric mushroom (*Amanita muscaria*). The smallest bite, although nutritious, can produce very strange behavior. Within an hour, the caribou leave the migratory procession and run awkwardly, shaking their heads and wagging their hindquarters from side to side, lagging behind the herd. In this state they are obviously in far greater danger of predation than normally. Jane Goodall saw a young golden jackal in the Ngorongoro Crater, in East Africa, eat a hallucinogenic mushroom, perhaps mistaking it for a more innocuous species. Ten minutes later he was rushing around in circles and charging, flat out, first at a Thomson's gazelle and then at a bull wildebeest. She never saw him eat it again.[19]

Not all narcotic nibbles are accidental, though. In the Canadian Rockies wild bighorn sheep will take great risks to reach small patches of an unidentified lichen that looks like thick yellow or green paint splashed on exposed rocks and boulders. Siegel found that local Indians use the lichen as a narcotic and the sheep grind their teeth to the gums in scraping it off the rocks.

Many species seek the narcotic effects of opium in its refined form. In India and Burma, mahouts put out opium balls to entice and capture wild elephants. In the wild, opium is much harder to come by, being available only from the opium poppy and only at certain times of the year. In those parts of Asia where opium poppies are grown commercially for the production of medicinal morphine, working water buffaloes that graze on cultivated opium poppies can become unwilling to work, and at the end of the season they show signs of opium withdrawal (restlessness, tremors, and convulsions). It is unlikely that the consumption is accidental; opium poppies are pungent, with a bitter taste, giving a clear indication of their toxicity. The buffaloes never eat enough to cause poisoning, but they appear to consume enough to provide pain relief (opium produces an indifference to pain rather than analgesia) and a feeling of well-being.

Siegel is confident that intoxication itself is the main drive behind such behavior — that animals seek drugs to find pleasure in a world full of unpleasurable sensations. In other words, they are self-medicating against life's discomforts. This thesis is supported by experiments in which animals in unpleasant surroundings are more likely to seek artificial drug-induced states of mind. Social isolation, for example, increases ethanol and opiate consumption in rats, and the stress of confinement increases the amount of morphine and fentanyl (painkillers) mice will voluntarily consume.[20] Many of the intoxicants animals consume are analgesic and, as we saw in Chapter 7, laboratory animals will readily self-medicate against pain. It may be that wild animals are similarly gaining pain relief from these plants. There are no documented observations of animals in overt pain seeking out narcotic plants, but then, as pain is often hidden, this behavior would be hard to spot in the field.

If we are to understand the full complexity of an animal's seeking intoxication, we need first to consider what motivates an animal to do anything. Hunger and thirst can be measured by the amount of effort an animal puts into obtaining food or water. What drives hunger or thirst is the disparity between uncomfortable sensations and comfortable sensations. As water levels in the body drop, an animal feels discomfort ("thirst," if you like) and seeks water to end the discomfort (quench the thirst) and incidentally rehydrate the body. When water is initially found, the act of drinking produces pleasurable sensations that reward the animal for having found water. But when sensors in the stomach and brain detect that water levels have risen sufficiently to rehydrate the body, the sensation of drinking becomes less pleasurable and the animal stops.

This model of motivation, with pleasure as the primary goal, is called the hedonic model. Any behavior motivated by pleasurable sensations (or by the removal of unpleasant ones) is open to abuse because the pleasure *itself* can become a goal. We see an example in our own eating habits. Sweet, high-energy foods give us transient sensations of pleasure to encourage us to seek enough energy, but the pleasurable sensations can lead us to overconsumption when sweet foods are readily available. Could it be that some intoxicants are pleasurable because they are necessary in small amounts?

Such an idea is supported by research suggesting that small quantities of cannabinoids (ingredients in cannabis) are essential for the normal functioning of the central nervous system in mammals. Mice deprived of cannabinoid receptors in their brains die suddenly without any preceding sign of illness. Andreas Zimmer of the National Institute of Mental Health says, "Without these natural cannabinoids and their receptors, we are more likely to suffer from some catastrophic lethal central nervous system failure." Researchers at Heinrich Heine University in Düsselfdorf, Germany, think cannabinoid receptors are implicated in schizophrenia and may be useful in treating the disease. Sufferers from schizophrenia reportedly self-medicate with cannabis. Perhaps the healthy functioning of the central nervous system requires a regular supply of natural psychoactive compounds.[21] If so, a natural predilection for intoxicants could be vulnerable to developing into addiction — just as energy seeking can contort into excessive eating or drunkenness. Having coevolved with traces of such compounds in our omnivorous diet, our physiology and that of the plants we eat are not only biochemically similar but interdependent.

## ANIMAL ADDICTS?

When an animal has become physiologically addicted or psychologically dependent on a drug, it will do almost anything to get it. Although we can readily induce both forms of addiction in laboratory animals, we know few examples of animals becoming addicted to natural drugs in the wild. Even a chapter entitled "Addictive Behavior in Free-ranging Animals" (in the book *Psychic Dependence*) starts by admitting that the title is a bit misleading owing to the lack of examples.[22] Wild koalas could perhaps be said to be addicted to their particular eucalyptus, in that when newly born they are happy to eat a range of plants, but soon focus on a small selection of eucalyptus species. If these are denied them, they will starve to death rather than eat other plants — even other eucalyptus.

Similarly, cattle seem to become addicted to some seriously toxic plants. Normally they avoid the burning acrid taste of buttercups in pasture, but young cattle turned out for the first time (and unfamiliar with *any* plants) will gorge on celery-leaved buttercups (*Ranunculus* sp.) and become very ill. If they recover and are turned out again on the same pasture, they will eat the same plants again with similar ill effects.[23] And those waxwings certainly return again and again to those fermented berries, with fatal results.

A remarkable example of wild addiction is cited by Siegel in his book *Intoxication: Life in Pursuit of Artificial Paradise*. One particular species of ant (*Lasius flavus*) lives in close relation with the Lomechusa beetle (named after an ancient Roman poisoner). In return for providing beetle larvae with food and care, the ants are allowed to lick an intoxicating secretion from the beetles' abdomens. Although temporarily disoriented and unstable on their legs, the ants become so addicted to the secretion that in times of danger they will move beetle larvae out of danger *before* rescuing their own. But addiction has its price and excessive consumption of this intoxicant can cause so much mania in an ant colony that reproduction stops and the entire colony collapses.

Luckily, in the wild, addictive intoxicants such as cocaine come in small packages that are not easily consumed in vast amounts. It would take several sacks full of leaves to provide a dose similar to that snorted by a human drug user, and during the time it would take to consume the leaves, the deadening effect in the mouth would deter most animals. In the wild, the taste and smell of many intoxicating plants restrict accidental overdosing. If too much is consumed, the natural responses of vomiting and diarrhea rapidly rid the body of these toxins. This lack of opportunity for abuse means that there has been little selective pressure to counter the propensity for addiction.

In contrast, the human habit of injecting artificially purified chemicals straight into the blood stream bypasses all these natural methods of avoidance, elimination, and detoxication. Artificially purified drugs are available year round to those who want them. No wonder our species is blighted by drug addiction! Our technology has overridden all the natural brakes that limit a predisposition to pleasure seeking. The social brake of legislation seems powerless against the unhealthy combination of natural desire and technological know-how.

Ironically, observing how and why animals become intoxicated could help us find effective treatment for our own drug addiction. Until recently, ibogaine (revealed by those animals in the Congo) was used in medicine as a general stimulant, but current research is focusing on ibogaine's ability to help reverse addiction to morphine and cocaine without itself being addictive.[24] New and effective analgesics, anesthetics, sedatives, stimulants, and mood modulators are always needed, and further studies of animal behavior in the wild may be one significant way to identify natural substances for investigation.

11

## PSYCHOLOGICAL ILLS

If you had roamed every continent
For thousands of years,
Coming to consider the globe your own private football,
And you were then confined to an open prison,
A tourist-infested allotment
in the suburbs of Nairobi,
On emergency rations,
You too might become unbalanced.

— Heathcote Williams, 1989

As I WRITE in the year 2000, a two-ton male elephant seal has been terrorizing local fishermen in the small town of Gisborne, New Zealand. Homer, as he has been named (after the overweight cartoon character in "The Simpsons"), is demolishing empty cars, rearing up and bringing down his massive weight, smashing the bumpers, fenders, and mirrors. The fishermen complain that Homer is blocking the boat ramp and stopping them from getting into the water, but to date no one has been brave enough to try and move him. This is not normal behavior for an elephant seal. Instead of fighting cars, he should be competing for females with other male elephant seals. It is not adaptive to waste time battling with inanimate objects; is he psychologically ill?

It is only relatively recently that scientists have accepted the likelihood that animals other than ourselves have minds — or at least mental states — and can therefore have healthy or unhealthy minds. The fear of anthropomorphizing animals becomes almost palpable when one starts to discuss psychological health with scientists, who are happier using terms such as aberrant or maladaptive behavior instead of mentally ill or psychologically disturbed. While these scientists are denying that animals have psychological states, laboratories around the world are using rats, guinea pigs, mice, dogs, and primates as models of human psychological states. There are autistic guinea pigs, neurotic mice, depressed dogs, stressed monkeys, and obsessive-compulsive rats. Many have been bred specifically for these traits; others have had them created by experimental conditions. The disturbed animals are then treated with the same pharmacological drugs with which we treat our own mental illness.

We cannot have it both ways: we cannot say animals have no minds, and then use them to model our own disturbed minds. Certainly, the minds of other species may be different (both qualitatively and quantitatively) from our own, but they are minds nonetheless.

Although the causes of psychological ill health cannot be definitively separated into genetic *or* environmental, nature *or* nurture, there are two distinct types of psychological ills to consider. One is the chronic condition in which an animal is predictably strange — what might be considered insanity in humans; the other is more acute, when an animal suffers a temporary episode of psychological ill health. Let us look first for signs of the former.

## NOT SO CRAZY AFTER ALL

Most seemingly crazy behavior of wild animals turns out to be not so crazy after all. One might be forgiven for thinking that rogue elephants that occasionally terrorize people by rampaging through villages, destroying huts and crops, killing farmers, and stalking individuals are suffering some form of insanity. If not insanity, then certainly a lack of survival skills, as they inevitably end up shot. Yet the reasons for their behavior are complex. When Tarquin Hall traveled to Assam in northern India in 1998 to record the hunt of a rogue elephant that had killed forty people, he found that as a youngster the elephant had been abused by an alcoholic owner and was now enduring unbearable pain from a rusty iron chain embedded in one leg.[1]

Occasionally a mature lion ruthlessly kills defenseless cubs, and may even eat them. However horrific this seems, the lion's behavior bears no relation to that of a human homicidal maniac who kills and eats his victims. The lion behavior occurs when a new male takes over a pride of females that already have cubs by previous males. By killing these cubs, the new lion ensures that the females come into estrus and all future offspring carry his own genetic material. The killing need not be cognitive, but merely a response to the unfamiliar odor cues of the cubs. The new cubs will carry *his* odor and therefore be protected by him. The killing behavior is adaptive for the adults, since a male that does not kill the cubs of other males will leave fewer offspring — he will be less fit. Natural selection makes no ethical assessments. Similar infanticide is seen in other group-living animals when a new male comes on the scene. Plains zebra and wild horses both experience infanticide and abortions when a new stallion takes over a herd.

One female common eider was seen adopting a clutch of eggs in the nest of another species, the gadwall. The behavior was clearly maladaptive, as she was wasting time and resources on eggs that were not her own (and not even related). She apparently did this because her own eggs had been eaten by gulls, and because she was disturbed by the large number of people who were around. Under unusual circumstances, she did something weird.[2]

Homer, the elephant seal, also turns out to have a reasonable excuse for his behavior. As a juvenile, he does not yet have access to his own harem, but he is driven to distraction by the spring molt, which makes his skin fester and his fur fall out. Elephant seals often become extremely irritable while molting, smashing and damaging things they come across in an attempt to allay the unbearable itching.

## SIGNS OF INSANITY

In the few cases in which wild animals have been studied over several generations, there have emerged very few but nonetheless intriguing signs of what might be called insanity. Jane Goodall described a female chimpanzee that, by any definition, was obviously disturbed. Passion (as she was called) repeatedly killed and ate the offspring of other chimpanzees in her group, and she was quite obviously feared and avoided by others. Even males showed little interest in her offers of sex, as her strange behavior left suitors confused. Perhaps the most poignant sign of her strangeness was her complete lack of regard for her own offspring. Her maternal neglect led to the death of several babies, and the only one who survived, Pom, became a highly disturbed adult herself.

Unlike other female chimpanzees, Passion had no close female friends and was uneasy and tense when others were nearby. Goodall is convinced that the lack of female companionship contributed to the poor survival of Passion's offspring. She was a cold mother, intolerant and brusque, seldom playing with her infant, particularly during the first two years. Pom was an anxious child, clinging and fearful, and at weaning time, after Passion had given birth to a male infant, the normally transient difficulties young chimpanzees experience at this period went on and on. While other weaned infants recover in a few months, Pom was lethargic and listless for three years. Later, as an adolescent, she joined her mother in killing the infants of other chimpanzees. (Cannibalism is rare in chimpanzees and even rarer among members of the same group.)[3]

It is uncertain whether Passion was disturbed as a result of genetic factors, or because of an inadequate upbringing, or if she was suffering from a physical ailment such as chronic pain or infection. Her behavior, consistently odd throughout her life, certainly reduced her fitness. Such examples of long-term psychological disturbance are extremely rare in the wild. Individuals with these sorts of problems are unlikely to survive for long.

On the other hand, most animals (including ourselves) may experience short-term blips in mental health. Furthermore, they appear to have strategies for dealing with psychological wear and tear in the same way that they deal with physical health threats. According to the veterinarian Beat Wechsler, animals have four main strategies for dealing with threatening situations: *escape* the problem, *remove* the problem, *search* for a solution, or *wait* until the problem goes away. These coping strategies are pivotal in avoiding psychological illness.[4]

Mental states such as anxiety, fear, excitement, and anger are necessary for reproduction and survival, but if they persist too long they can be counterproductive. An antelope that is fearful of an approaching lion may live another day, but an antelope stuck in a *constant* state of anxiety will not be able to fulfill all its other behavioral needs. Somehow, animals appear to balance their psychological states. Some naturally disperse as the situation changes. For the rest, as we shall see, there are many ways in which animals (mammals in particular) change their behavior in order to maintain their psychological health.

## COPING IN THE WILD

Stress is defined by physiological changes that occur in response to an aversive situation. The immediate response is that the body mobilizes energy for flight or fight. Long-term building and repair projects, such as growth and immunity, are deferred until the emergency has passed. Pain is temporarily blunted and memory sharpened. There is a rise in glucocorticoid hormones, making the animal anxious and ready for action, but also distracted and unable to concentrate on other important activities such as reproduction and feeding. This is effective in dealing with short-term emergencies, but prolonged stress responses are harmful. Exhaustion sets in and suppression of the immune system exposes the animal to infectious diseases. The mental state therefore needs to be brought back into balance as soon as possible.[5]

One way animals reduce their anxiety levels is through grooming, hugging, and stroking themselves. Both wild and domesticated animals defer feeding in favor of self-grooming when they are subjected to physical or emotional stress. As we saw in Chapter 7, baboon society is violent and tense. Fighting is common and serious injury is frequent. Living in such an aggressive society has produced some interesting coping strategies. Soon after a conflict, the combatants scratch, groom, or shake themselves and yawn to release tension. If those measures are ineffective, they often seek some kind of reconciliation, after which their need for these so-called self-directed behaviors reduces. Most primates try to make up after fighting — or at least make sure the fight is over.[6]

Sometimes just being near a more powerful peer can produce great anxiety. When a high-ranking female baboon is within 5 meters, a low-ranking female will start to shake, groom, touch herself, and yawn, until the senior female moves away. The behavior is so predictable that it can be used as a measure of an individual's anxiety, or even to predict the status of individuals passing by, just as we might use the reactions of people present to gauge the importance of an individual entering the room.

In experiments in which levels of stress hormones have been monitored, it is apparent that these self-directed behaviors are very calming. When a high-ranking bushbaby approaches a low-ranking bushbaby and sits nearby, the rise in its blood cortisol levels shows that the situation is stressful for the lower-ranking animal. The anxious bushbaby starts to groom herself — more specifically, to rub her feet and chest. This behavior coincides with a decrease in stress hormone levels. What is more, individuals who rub their chest and feet cope better with the stress of a novel environment than those that do not. One explanation is that secretions from scent glands in the chest are reassuring to bushbabies, turning an unfamiliar smell into a familiar one. Scent marking is considered an effective way for many prosimians (early primates such as lemurs and bushbabies) to reduce the physiological arousal of stressful situations.[7]

Mice and rats can self-medicate against emotional stress. In a particularly unpleasant experiment, mice were exposed to two different types of stress. One group received electric shocks to the feet (described as "acute physical stress," but known to most people as "pain"); the other group was forced to witness *another* mouse getting a foot shock (described as "acute emotional stress"). Both groups had free access to mind-numbing and pain-numbing morphine, but only the mice exposed to emotional stress self-administered the morphine.[8]

This experiment shows two things: first, that the emotional stress was unpleasant; second, that the emotional stress (but not the physical stress) made the mice more receptive to the rewarding action of morphine. The researchers suggest that whereas the mice can associate physical pain with an obvious source and therefore cope with the stress, the emotionally stressed mice have no way of knowing what is happening and are therefore less able to cope with the situation. A similar effect is seen in emotionally stressed rats and cocaine self-administration.[9]

Laboratory rats are also able to use biofeedback to calm themselves. Scientists in the Ukraine found that stressed rats learned to self-administer strobe lighting at certain frequencies that changed electrical activity in the brain, and thereby calmed heart rhythm and lowered blood pressure. In this way the rats ingeniously calmed themselves, reducing the likelihood of heart attack.[10] A feeling of anxiety is assuredly unpleasant, and it is the animals' desire to *feel* better that drives the coping strategies, including self-medication.

The degree to which wild animals regulate their mental states with diet has not been researched, yet diet certainly can affect mood. There are traces of valium-like compounds in everyday foods such as potatoes and fruits, and some people combat anxiety and depression with calming, soothing foods.[11] In Chapter 15 we will hear a fascinating anecdotal report from Apenheul Zoo in the Netherlands about monkeys that sought out a calming herb, valerian, after stressful fights. Some forms of obsessive and compulsive behavior in domestic dogs can be treated with diet, and odd behaviors of confined competition horses can be eradicated by a change of diet. By exploring this link between dietary choice and animal behavior in the field, we might be able to uncover a range of safe mood modifiers for ourselves.

## SHAKING OFF TRAUMA

In recent decades there has been much talk of posttraumatic stress syndrome (PTSS) — a state of psychological ill health that arises in humans after extreme emotional trauma. Psychologist Peter Levine, who has provided therapy for this disorder for thirty years, made a real breakthrough when he started to think about how animals in the wild cope with trauma. Wild animals do not appear to suffer from the debilitating effects of PTSS even though they frequently deal with life-threatening situations. When faced with imminent danger, many species freeze into what Levine describes as "an altered state of consciousness," in which no pain is felt. Humans unfortunate enough to have been mauled by lions tell how they were frozen to the spot but felt no pain — a phenomenon known as stress-induced analgesia. The freeze response in animals too is thought to be adaptive, as many predators need movement on the part of their prey to be triggered into killing.[12] The endorphins (internal painkillers) rushing around in the frozen animal's body contribute to its survival. Shock from intense pain would hinder opportunistic escape.

Levine suggests that some people become stuck in the freeze response, and it is this response that causes posttraumatic stress syndrome. Animals, he says, avoid this situation by physically shaking themselves out of a frozen state. Certainly many species do shake after stressful events — and not with just a tremor. Apes, monkeys, cats, and dogs give themselves a full body shake in response to intense stress, both acute and chronic.[13]

If an animal can neither escape nor remove an aversive situation, it is not adaptive to attempt the same coping strategies over and over. Often, when ethologists first start to watch wild animals (even from afar), the animals simply go to sleep. Goodall makes the point that this is not necessarily a sign that they are relaxed but, more often, that they are extremely uneasy. If a captive chimpanzee is presented with a problem it cannot solve, it will curl up on the bare floor and go to sleep. It is a form of escape — a reaction commonly reported by hunters, too. Trapped animals go into a deep sleep, almost as if comatose. Humans may also perform "cut-off behavior" (closing the eyes, remaining still, not interacting) as a defensive strategy in response to an inescapable threat.[14]

The strategy of waiting for change to come about is often described as apathetic behavior or learned helplessness. In its extreme, we know this as depression, characterized by listlessness, lack of appetite, lethargy, lack of interest in others or events, reduction in personal hygiene, and lack of responsiveness. Goodall has described the last days of a severely depressed chimpanzee: "After the death of her infant, Melissa seemed to lose the will to live. She had been thin before, now she became emaciated for she ate almost nothing. Often she did not leave her nest until after ten in the morning and sometimes she went to bed as early as four o'clock. During the hours in between she made at least one day nest where she lay, often staring vacantly up through the leaves, for hours at a time."[15]

In wild chimpanzees, depression is a common response to weaning and the loss of a parent, friend, or offspring. If the depression gets so deep that the animal loses the will to live, recovery seems to depend on many factors, the most obvious being the support of other group members. In the laboratory, chimpanzees become depressed when faced with a survival problem that has no apparent solution. Depression then can be seen as an appropriate coping strategy for a situation where there is nothing to do but wait for things to change. Proponents of Darwinian medicine also see depression as a way of keeping an animal from continuing to perform unsuccessful (maladaptive) actions. Again, depression can be a *coping* strategy rather than a psychological problem itself.[16]

Humans often tend to avoid interacting with mentally ill people, and in this respect we are no different from other animals. The most common reaction to abnormal behavior among animals is social exclusion. In terms of Wechsler's coping mechanisms, this can be seen as escape from and removal of the threatening object. Psychologically ill animals send out confusing signals to their fellows. They may not follow the behavioral norms that keep aggression within bounds or that reduce risk of disease. As strange behavior can be caused by infectious disease, shunning those that act strangely may be one way of protecting oneself against infection as well as unpredictable aggression.

The examples above, and many others that have accrued over two decades, allow us to conclude that animals have many behavioral strategies for rebalancing psychological homeostasis. They have a range of self-calming actions, including an ability to turn off if life gets too difficult. They shake themselves, and they may seek reconciliation. They are even able to reduce the uncomfortable sensations of emotional stress with mood-altering drugs.

## BORN FREE

Natural selection has not eliminated in wild animals the disturbed behavior that we see in ourselves and in our pets. Captivity does strange things to wild animals, species that have not been selectively bred for tolerance of human contact or confinement. When originally captured, they may attempt the first coping strategy by trying furiously to escape, injuring themselves or even dying in the process. Alternatively, they may fall into a catatonic state of complete shut-down.

The most common sign of disturbed behavior in captive animals is some kind of stereotypical movement such as rocking, weaving, or pacing. (By definition, stereotypical behaviors are not seen in a natural setting; they emerge only in captivity, particularly in carnivores and primates.) A polar bear paces up and down his enclosure, swinging his head regularly and hypnotically from side to side, then at the end of the loop rises up on his forelegs, swings around, and goes back the same way. Each loop is the same, over and over. An elephant sways from side to side, rubbing his chest on the iron bars of his enclosure. He rubs in exactly the same spot, so that the metal is worn smooth by his movements.

There are several theories about stereotypical behaviors: they are condensed versions of behaviors that the animals would perform in the wild; they are some kind of coping strategy, such as searching for solutions to confinement; they may even be self-medication, as the repetitive movements somehow soothe frustration. Universally, though, it is acknowledged that stereotypical actions are a distinct indication of poor psychological health.

Another anomaly that turns up in captivity is an eating disorder. Voluntary regurgitation and reingestion of food (akin to bulimia in humans) occur in up to 65 percent of captive gorillas, yet this behavior has never been seen in the wild. The disorder has been variously blamed on the unsatisfying nature of their diet, stress, early trauma, maternal neglect, or intense boredom. Other signs of psychological imbalance are a heightened aggression that goes far beyond that seen in the wild. Confined quails and pheasants peck each other to death, yet they rub along without extreme aggression in the wild. Captive capybara take to killing their own young.[17] As a plant eater, an elephant only kills when faced with extreme danger. Yet in the time it took to write this chapter, two cases of elephant homicide became news. An elephant in Port Lyme Zoo, Kent, England, crushed his keeper against the walls of his prison. A few weeks later, a performing elephant in Thailand went berserk, threw down his keeper, and ran to the audience, where he killed a young girl and injured many in the crowd. A wild animal that is not coping with its confinement is a dangerous animal. Often such overly aggressive animals are kept in isolation, but then they can turn on themselves. Self-abuse is unrecorded in the wild, but captive parrots pluck their own feathers, and mink chew their fur out by the roots.[18] These examples show that wild animals are susceptible to a range of familiar psychological disturbances, but that these are often seen only in captivity when natural coping strategies are denied them.

Among humans, depression is, according to the World Health Organization, the single most common cause of disability throughout the world. More than two million people in the United Kingdom suffer from obsessive compulsive disorder, a kind of permanent anxiety loop that can leave little room for normal life. Almost 10 percent of Americans have, or have had, a classifiable anxiety disorder. Sadly, self-inflicted damage is the leading cause of injury for women in the world. With our neuroses, self-abuse, eating disorders, addictions, and depression, humans look frighteningly like captive wild animals prevented from invoking their natural coping strategies. Desmond Morris concludes that cities are not concrete jungles, but human zoos: "The comparison we make is not between the city-dweller and the wild animal, but between the city-dweller and the captive animal."[19]

These comparisons between captive and free-ranging wild animals tell us that we should look to our lifestyles — our *coping strategies,* social interactions, and diets to improve our mental health. By acknowledging that psychological health needs to be actively managed, we may learn how to find both health *and* happiness.

12

✄ ✄ ✄ ✄

# FAMILY PLANNING

A small family is soon provided for.
— English proverb

MUCH HUMAN MISERY is caused by reproduction — too much, too little, too soon, too late, wrong sex, wrong partner. As a result, women have always attempted to control fertility — with or without the approval of their religious leaders. Occasionally the consequences for their health are disastrous. In the wild, animals also have to manipulate their reproductive health. There is a delicate balance to be had between maximizing reproduction and minimizing the drain on resources. For a female, investing time and energy in reproduction at the wrong time, or with the wrong male, or in offspring of the wrong sex, can greatly impair her health and reduce her lifetime fitness.

Learning how wild animals maintain their reproductive health is vital to conservation, farming, pest control, and even our own reproductive health. Captive breeding programs of endangered species are plagued with inexplicable reproductive problems. Pandas rarely conceive. Maned wolves may breed, but most of their offspring die soon after birth. Elephants commonly fail to sustain an interest in sex, even with a receptive new mate, and the testes of many captive gorillas shrink and shrivel until useless. The stress of confinement is only one reason for these difficulties. Major players in animal reproduction are those plant secondary compounds that we have been discussing in such detail.

In 1946, ewes in Australia became permanently sterile after grazing on a certain type of clover (*Trifolium subterraneum*). The clover was later found to be rich in an isoflavone that disrupts reproduction.[1] For reasons not yet fully understood, many plants contain hormones that mimic the actions of animal hormones in both male and female vertebrates. So far, over twenty phytoestrogens (plant compounds that mimic estrogen, the female reproductive hormone) have been identified in more than three hundred plant species. Even everyday foods have estrogenic activity: licorice, fenugreek, rhubarb, garlic, and soybeans, to name a few. Although phytoestrogens are capable of turning reproduction on or off depending on the potency and timing of consumption, in birds and mammals they usually depress fertility. They are of growing interest because in addition to their influence on fertility, they are protective against many diseases, such as cancer of the breast, colon, and prostate.[2]

Progesterone, the main ingredient of the first modern contraceptive, came from the plant compound diosgenin, extracted from wild Mexican yams (*Dioscorea mexicana*). This compound is one of a handful of steroidal saponins that mammals use to make their reproductive hormones. Until 1970 diosgenin was the sole source of steroidal contraceptives, and although it still provides almost half of the raw material for steroid production, new sources are constantly being sought. The search so far has uncovered stigmasterol and sitosterol in soya (*Glycine* sp.), hecogenin in sisal (*Agave* sp.), sarsapogenin from *Yucca* and *Smilax* sp., and solasodin in *Solanum* species. It is evident that plants provide many compounds capable of influencing animal reproduction.

Artificial chemicals can also mimic estrogens. In the industrial world, male sperm quality has been dropping during the last four decades, while estrogen-related diseases have been rising. Many species of wildlife are showing signs of feminization, with severe disruption of reproduction and development. It is thought that increasing levels of unnatural estrogen mimics, mainly from waste products of the plastics industry, are responsible, but we will have to learn more in order to arrest the health-damaging effects of this hormonal pollution.[3] Paradoxically, then, although our historical interest in reproductive control has revolved around a desire to block pregnancy, we now urgently need to unblock the blockers that may have inadvertently arisen from our industrial waste. Observing how animal reproduction is modulated by phytoestrogens in the wild is a major part of that quest.

## FEEDING AND BREEDING

All animal reproduction is ultimately reliant on plant chemistry, but some species are more directly reliant than others. Butterflies, for example, need the alkaloids in *Heliotropium*, and cucumber beetles need the cucurbitacins in cucurbits, to produce reproductive hormones. Without them, reproduction is impaired.[4]

When low rainfall creates high concentrations of phytoestrogens in their legume diet, the number of eggs and hatchlings produced by the California quail falls. So in drought years, reproduction fortuitously goes down along with food supply. Similarly, when phenol levels are high in their cypress diet, Japanese voles have fewer offspring per litter.[5]

Temperate climes usually engender dramatic changes in seasonal food supply that require clearly defined breeding seasons. Red deer exposed to the harsh climatic alterations of the Scottish highlands have their breeding turned on and off by changing day length, as this adequately reflects food supply for their young. But day length is not always the best indicator of food supply, nor the most effective cue for reproduction. Montane voles, small herbivorous mammals native to North American grassland, live no more than one year and one breeding season. The flush of grass on which they feed varies from year to year depending on the thaw of mountain snows, so timing their breeding via day length would not be sufficiently accurate. Instead, reproduction is switched on and off directly by their diet. As their salt grass (*Distichlis stricta*) diet flowers, just before dying back, it releases two phenolic acids in high concentrations. These phenols cause reproductive effort to cease as the voles' food supply dries up. The following spring, another grass compound stimulates the onset of breeding. In this way, reproduction is accurately synchronized with food availability.[6]

The reproduction of many other herbivorous mammals, such as rabbits, kangaroos, and desert rodents, is also fine-tuned by their diet. In these examples, control over reproduction is incidental to feeding; in other seasonal breeders this may not be the case. Among gelada baboons in Ethiopia, few pregnancies occur in the dry season when females (but not males) eat *Trifolium* plants related to the clover that sterilized ewes in Australia. This suggests that the females may actively or fortuitously reduce their fertility during unsuitable conditions.[7]

Diet also affects the timing of reproduction in some tropical mammals that have no obvious seasonal breeding pattern. For free-ranging vervet monkeys in Kenya, reproductive behavior is related to the availability and ingestion of mimosa flowers (*Acacia elatior*). If these plants bloom early, the monkeys come into season early. If flowering is extended, so is the breeding season. The relationship holds true even for individual monkeys. The flowers contain phytoestrogens, and the monkeys eat enough of them to affect reproduction; whether they actively modulate their reproduction is not yet known.[8]

One species of monkey does seem to be particularly active in controlling its own fertility. In pockets of isolated forest in Brazil live the last of the world's rarest primates, the muriquis. In preparation for breeding, these monkeys go out of their way to eat the leaves of *Apuleia leiocarpa* and *Platypodium elegans*, which are high in protein. Then, at the beginning of the rainy season, the optimal time for pregnancy, the monkeys make a particular effort to find and eat the fruit of the "monkey ear," *Enterolobium contortisiliquum*, which contains stigmasterol, a steroid used by the body to synthesize progesterone. Such a compound is capable of triggering reproduction.[9]

According to Chinese folklore, the cordyceps or caterpillar fungus (*Cordyceps sinensis*) was first discovered by yak herders, who noticed that their yaks readily ate it at the beginning of the breeding season. When herdsmen tried eating some themselves, they became more jolly, agile, and strong, and claimed that with prolonged use several chronic immune and respiratory illnesses vanished. The fungus was even said to have antiaging properties. Although it was reputed to be effective, it was also rare. Consequently, it became extremely expensive and eventually was reserved for the emperor's physicians only.

In recent decades the Chinese have learned to cultivate this fungus commercially, so that ordinary mortals can now enjoy its numerous benefits. The National Health Ministry of China admitted that it gave *Cordyceps* daily to the women's track and field team that set nine world records in 1993. Modern phytochemical research shows that this fungus contains naphthoquinones, antitumor sterols, and a newly analyzed ingredient that has anti-HIV activity. Because of its ability to reverse the loss of sexual drive in humans, the ethnobotanist Mark Plotkin has dubbed it "fungal Viagra."[10] It is not hard to imagine that animals might find such a food pleasurable.

Another Chinese legend tells of a goat herder who noticed that his male goats became sexually aroused after feeding on a particular plant (*Epimedium sagittatum*). This plant is one of the most potent herbs used in Chinese medicine for enhancing male potency and is known colloquially as horny-goat weed.[11]

While humans continue to debate the rights of the unborn child, other species routinely terminate unwanted pregnancies. The timing of births is critical in the wild — depending on resources and environment — and animals often find themselves pregnant at a less-than-perfect moment. Goats, rabbits, cats, marmots, rodents, and horses simply resorb very young unwanted fetuses into their bodies, wasting no resources in the process. Resorptions such as these can be triggered by a range of social, psychological, and physiological mechanisms, but a common cause is the arrival of a new male on the scene. Indeed, the odor of a strange male is sufficient to trigger resorption in laboratory rodents, caged rabbits, and wild marmots.[12] And if odor can terminate a pregnancy, it is certainly possible that a change of diet can too. Joan Garey noticed that wild female chimpanzees occasionally feed on the same plants local people use to abort fetuses (*Combretum* and *Ziziphus* leaves), although it is not known whether the chimpanzees were pregnant before or after consumption of these plants.

## DESIGNER BABIES

Recent technological advances in human reproductive biology have made it possible to choose the sex of our children. Yet this is not the first time humans have manipulated the sex ratios of their offspring. Throughout history, in societies where boys bring greater economic return than girls, the occasional result is the killing of female babies.[13] In the wild, the costs and benefits of sons versus daughters also vary depending on a species' social organization (although the parents may not always agree on the value of each sex). During the upbringing of great reed warblers, both fathers and mothers bring food to the chicks, but while mothers devote an equal amount of time to sons and daughters, fathers invest more time in feeding their sons. Male great reed warblers are thereby preferentially investing in sons.[14]

The reason offspring of different sexes have different values for animal parents is that selection favors parental strategies that produce not only children but grandchildren and great-grandchildren, and sons and daughters have different chances of doing this. Red deer have polygynous harems (one male, several females), so that very few males breed at any one time, but most females get an opportunity. Having daughters is therefore a safer "grandchild bet" than having a son. What is more, sons need to grow big and strong to succeed, and thus they need more resources. They are both high investment and high risk. But if a son *does* win a harem, he can father more offspring than a daughter can. The investment risk also varies according to the mother's social status. Because the future prospects of offspring are influenced by the mother's rank, high-ranking hinds can invest in sons with less risk, as their sons are more likely to gain a harem. A twenty-year study of red deer on the isle of Rhum showed that high-ranking hinds do indeed produce more sons, while lower-ranking hinds produce an excess of daughters.[15] How this occurs is not yet known.

Food availability can have a powerful influence on sex ratio. Leaf cutter bees produce more females when resources are ample, and in eusocial insects such as the ant, abundant food produces a bias toward females, whereas a shortage of food produces a bias toward males. And house mice on low-calorie diets give birth to fewer males.[16]

Food quality may play a role as well. When worms of the species *Dinophilus gyrociliatus* were fed cereals, rather than spinach, they gave birth to more males. In Kaimanawa horses, sex ratio is determined by the health of the mother at conception. Mares in better condition have more foals of the more expensive sex — males.[17]

Sex ratios in our closest relatives, the primates, appear to be influenced by the type of food they eat. Primates usually have single births spread well apart, each one representing a substantial investment on the parents' part. It is therefore more cost effective to manipulate sex ratio at conception. Twenty-two years of birth records suggest that some female mantled howler monkeys may be doing exactly this. Kenneth Glander found that female howler monkeys in Costa Rica give birth repeatedly to offspring of one sex more than another (with statistically significant consistency). How could this happen? In primates, sex is determined by sperm from the father. If an X chromosome fertilizes the female's egg, a female is conceived; if a Y chromosome fertilizes the egg, a male is conceived. These X and Y chromosomes respond differently to the electric potential in the vagina, because Y-carrying sperm carry a negative charge; X-carrying sperm, a positive charge. If the electric potential of the vagina can be altered by diet, Glander argues that primates could affect the sex ratio of their offspring.[18]

In addition, diet selection may influence the frequency of twinning in primates. Scientists from New York University noticed that free-ranging rhesus monkeys on Morgan Island, South Carolina, had an unusually high incidence of twinning, and that they also eat roots of the *Smilax bona-nox*. Smilax plants contain steroidal compounds known to be estrogenic; this plant is therefore suspected of affecting the incidence of twins.[19]

A recent review of the evidence for sex-ratio manipulation in primates revealed that it results from a combination of psychological, physiological, and environmental factors.[20] Although it is accepted that many animals show adaptive sex-ratio manipulation, it is too early to say precisely what role behavioral strategies contribute. However, when push comes to shove, one bird certainly manipulates the ratio directly. The zebra finch, a tiny bird that usually gets only one attempt at breeding, is under intense pressure to get the sex ratio of its clutch absolutely right. Rebecca Kilner found that the parents selectively *push* the unwanted sex out of the nest soon after hatching.[21]

## A HEALTHY PREGNANCY

For humans, early pregnancy can bring nausea, vomiting, and a pronounced change in dietary preferences. Smells and tastes can become so acute that it is hard even to be near certain foods. As the evolutionary biologist Margie Profet explains, since the delicate developing fetus can be harmed by toxins in the mother's diet, forgoing certain foods in early pregnancy is highly adaptive. Other mammals also appear to feel queasy during early pregnancy. Domestic dogs go off their feed and sometimes vomit, and American elk slow their grazing. Wild chimpanzees too go off their food. Intriguingly, Joan Garey has seen wild chimpanzees eat small amounts of acacias, hibiscus, smilax, *Alcornea cordifolia*, and *Celtis africana*, all used by local people to treat stomach upsets and morning sickness.[22]

Slightly later in pregnancy, it is common for women to crave particular, often unusual foods — or nonfoods such as earth, coal, or charcoal. This preference may reflect a nutritional craving for minerals such as iron, but it could also be an adaptive strategy for adsorbing damaging toxins. In the wild, we see evidence of pregnant animals clearing out toxins, parasites, and pathogens. Worm loads are lower in pregnant and lactating female primates than in nonpregnant females.[23] Self-medication may play a role here. As we saw earlier, female chimpanzees swallow hairy worm-scouring leaves more frequently than males, and they chew more worm-killing bitter-pith than males. Similarly, female macaques eat more gut-lining and toxin-absorbing clay than males.

In later pregnancy, eating generally increases as mothers need to provide extra nutrients for growing fetuses. Heavily pregnant American elk almost double the speed of their grazing, and they graze for much longer periods each day. Near birth more acute discomfort may be present. Anyone who has carried a child to full term will know just how uncomfortable those last few weeks can be, and it is not surprising to find animals attempting to relieve these unpleasant sensations. When Holly Dublin of the World Wildlife Fund followed a pregnant African elephant for more than a year, the elephant generally ate a predictable diet, roaming about 5 kilometers each day.

Then, at the end of her long gestation, she suddenly walked 27 kilometers in one day. She headed to a tree of the Boraginacea family and ate it — leaves, trunk, and all! Four days later, she gave birth to a healthy calf. A Tongwe elder, Babu Kalunde, reported having seen a heavily pregnant elephant do the same in a remote part of eastern Africa several years earlier. Dublin found that Kenyan women brew a tea from the leaves of this tree to induce labor. The tree is in the same plant family as borage (*Borago officinalis*), used by European herbalists to improve the flow of milk, and stoneseed, employed as a contraceptive by native Indians.[24]

Jane Pointer, a retired organic farmer in southern England, regularly saw her heavily pregnant cows eat flowering dock down to the ground the day before, or the day of, giving birth. Dock (*Rumex* sp.), which tastes bitter, generally contains anthroquinone glycosides and tannins and is used by herbalists for a congested liver and constipation — two common symptoms of late pregnancy. Whether this plant contains something even more interesting during flowering has yet to be established.

The production of milk requires extra nutrients, especially calcium and protein, and as we saw in Chapter 3, lactating mammals are able to find and ingest greater amounts of calcium than nonlactating females. Lactation in ring-tailed lemurs of Madagascar coincides with the availability of high-quality, high-protein food in the hot wet season, and they eat a wider variety of species to avoid overloading their detoxication system with one particular toxin. They concentrate their feeding on young leaves and buds — high in protein, low in toxins — avoiding the fruits that are eaten by nonlactating lemurs. They rest much more and, not surprisingly, spend more time on the ground than in the trees. In this way, they maximize their health and the health of their offspring.[25]

Lactation can bring its own health hazards, as sensitive breast tissue is opened to the outside world and sucked on by hungry mouths. For those mammals that can reach their nipples, licking them with antiseptic saliva is an excellent way to heal sores and avoid breast infection. Newborn rats will refuse to attach to their mother's nipples unless the nipples are covered in the mother's antiseptic saliva. This fussiness not only ensures that the young rat gets a boost of its mother's helpful microorganisms, but also protects it from harmful bacteria on the nipples.

Even courtship is affected by an animal's desire to avoid disease and pass on genetic tendencies for superior health to its offspring. *Looking* healthy is important for acquiring a mate as well as for maintaining social position. The male peacock is the best-known demonstrator of health. The fact that he can survive and thrive despite the burden of those long, elaborate feathers means that he is strong and healthy and free of feather-damaging pathogens. Any lapse in health will be visible in lackluster feathers, so females can select healthy males by selecting displays of healthy feathers. Experiments show that quite subtle health defects can be discerned by animals. Female grain beetles can smell which males are infected with tapeworms and prefer to mate with uninfected males. Such sexual selection, along with natural selection, is one of the two main forces of evolution.[26]

Among some butterflies, moths, and beetles, the female appears to select a mate based on his ability to provide her young with pharmacological health care. The male hands over plant toxins at the time of mating as a nuptial gift, and the female assesses her potential mate by the strength of the smell the gift emits. She then not only uses these compounds for her own protection but inserts them into her eggs to protect them from infection and predation.[27] Copulation itself is hazardous, as its distractions leave animals vulnerable to predation. Even in this circumstance, insects use plant toxins to protect themselves. Male scarlet-bodied wasp moths harvest alkaloids from plants such as dog fennel, and store them in abdominal pouches containing a mass of intricate fibers. The male discharges this toxic web over his mate, protecting them both from predation while they copulate for up to nine hours!

In the wild, unprotected homosexual, heterosexual, and multiple matings are common, and although it is difficult to gauge the frequency of sexually transmitted diseases among wild animals, they must present health hazards similar to those of other forms of infectious disease. One way rats lessen this risk is by licking their genitals after intercourse. This behavior effectively reduces transmission of such diseases because their saliva contains potent antimicrobial compounds.[28]

It is too early to estimate the degree to which dietary control of reproduction is merely incidental to feeding. As reproduction is so thoroughly interwoven with the chemical ecology of each species, essential dietary compounds may be needed for successful breeding and care of offspring. We need to learn more about how diet interacts with our own reproductive biology, too. What are those cravings in pregnancy trying to tell us? Is morning sickness a blessing rather than a curse? Is fasting natural in early pregnancy? Questions such as these, along with the increases in estrogen-related cancers and infertility, mean that observational studies of wild reproductive health are perhaps more essential today than ever before.

# 13

※ ※ ※ ※

## FACING THE INEVITABLE

> By medicine life may be prolonged, yet death will seize the
> doctor too.
> — William Shakespeare, 1609

IN GOMBE, wild chimpanzees give birth to their first offspring when
they are between twelve and twenty years of age, remain fertile into
their mid-thirties, and then survive only another three to nine years.
Old age creeps in during their late thirties and early forties, with a
general decline in activity, loss of coat condition, graying and thin-
ning hair, loose and worn teeth, and withdrawal from social interactions. Roughly 12 percent of the population are old at any one time,
and during thirty years of study only three females and six males
lived to become "ancient looking."[1]

For most animals in the wild, aging is a rare achievement. When
old age comes, it brings specific health challenges — simply staying
alive can present difficulties, as failing eyesight and weakened muscles reduce one's ability to stay out of trouble. Even getting around
can be hazardous, as this account of an elderly spotted hyena in the
Ngorogoro crater illustrates.

> The old male, Nelson, is blind in his right eye and his ears are tattered
> from endless bickerings over food and females . . . He walks with his neck
> held rather stiffly and his head twisted to one side so that the good eye
> points ahead . . . Often, too, he trips over tufts of grass or other irregulari-
> ties in the ground, and once, as he walked past a den, he stumbled right at
> the edge and vanished down the deep burrow head first.[2]

We do not know how commonly wild animals are affected by diseases we humans associate with old age — arthritis, cancer, dementia, depression, and heart disease. Certainly some signs of arthritis occur in the wild. A few of the six hundred wolf and coyote carcasses examined in Canada showed signs of degenerative joint disease, and wild gorillas and feral horses both show a tendency to osteoarthritis as they get older. Fertility declines in mammals, and old age is usually a postreproductive period in life. However, the current scourges of human aging appear to be rare in the wild. Although aging apes carry the genes associated with Alzheimer's disease in humans and accumulate the same insoluble lumps of protein (plaques), they do not show any indication of dementia, even in advanced years. The only mammals known to have levels of cancer similar to our own are the last 650 beluga whales confined to the heavily polluted industrial waters of the Saint Lawrence estuary in eastern Canada.[3]

## WILD AGING

Why organisms die, and why they age, are different questions with different answers. For example, species such as sea anemones are reported not to age, but they do die. Immortality, it seems, is not adaptive. But all land vertebrates and insects undergo senescence — the process of bodily deterioration that brings an increasing susceptibility to disease, a decreasing ability to repair damage, and a host of age-related health hazards.

The hard fact is that life span is directly linked to reproduction. Once reproductive peak has been passed, aging and death inevitably follow. Delay breeding to later ages for a few generations, and you can extend longevity. By allowing fruit flies to reproduce only at an advanced age, scientists have managed to double the natural life span of the flies. Aging could be the price paid for reproductive advantage in earlier life.

An example, used by Randolph Nesse and George Williams in their book *Why We Get Sick*, involves an imaginary gene that changes calcium metabolism so that bone heals faster, but also causes slow and steady calcium deposition in the arteries. Such a gene could be selected in the wild because an individual would benefit from the advantages of faster healing in youth; but wear and tear, predation, and accident would probably keep it from living long enough to suffer the disadvantages of hardened arteries in old age.[4]

Life span can be extended by diet — or, more accurately, by a lack of food. The lifetime of laboratory mice and rats can be increased by 30 percent simply by restricting calorie intake to near-starvation levels. An extreme calorie restriction switches off reproduction (in the same way that it causes human gymnasts and anorexics to cease menstruating), and it is this pause in reproduction that extends life, as the animals are held in limbo waiting for reproduction to start.[5]

Contrary to popular belief, humans do not live longer than they used to. More of us survive to an older age, but the maximum human life span — about 120 years — has not increased. Whereas in fifteenth-century England the average life expectancy at birth was as low as twenty-three years, now it is seventy-six years. However, longevity is not necessarily a reflection of good health. The World Health Organization has devised a new way of measuring health that takes disease and disability into account. It is called the Disability Adjusted Life Expectancy (DALE) score and effectively measures the length of *healthy* life rather than longevity alone. The United States ranked twenty-fourth in the health ratings, with Americans reaching an average of seventy years before being disabled by ill health. Japan was a clear winner, with people reaching an average of 74.5 years.[6]

Reasons for the differences relate primarily to diet and smoking. As more and more of us survive into old age, the health strategies of successful aging will become more and more valuable, and we can learn much from wild animals. Because so few of them survive into the later years, we should not expect to find that they have evolved many behavioral strategies *specifically* for dealing with the health hazards of old age. However, as they strive to maintain health continuously throughout life, some of the behavioral changes seen in old animals may be useful and transferable to our own lives. Furthermore, we have much to learn about ourselves from the ways other species deal with death and dying.

Successful aging calls for changes in behavior — particularly dietary modifications. Elephants get six sets of teeth, and although that would seem sufficient to those of us who only get two, six are not always enough. Elderly elephants therefore hang out in swamps and water-filled areas, where vegetation is softer and easier to chew. Like an elderly human patient on liquidized food, elderly elephants chew soft swampy plants and drink plenty of water.

Folklore suggests that we might find valuable antiaging compounds by observing what elderly animals eat. In China, herdsmen noticed that elderly deer (not those that were younger) nibbled at the bitter and astringent bark and roots of the fleeceflower (*Polygonum multiflorum*), or fo-ti plant, and when this plant was analyzed, Chinese herbalists found it had many useful properties for elderly patients. It is utilized by Eastern and Western herbalists as a tonic to maintain youthful vigor and increase energy, and it is reputed to reduce hypertension, cholesterol levels, and coronary heart disease.[7]

A group of natural compounds currently under research for their beneficial effects on aging are the antioxidants such as vitamins A and C, minerals such as selenium, coenzymes, and the hormone melatonin. One theory of what causes aging is that, throughout life, cumulative damage is accrued through processes (such as the metabolism of food) which release free radicals that harm other cell structures, including DNA. Antioxidants soak up free radicals; thus a diet rich in antioxidants could slow the aging process. The diets of wild plant eaters generally contain numerous natural antioxidants. Blueberries, for example, contain polyphenolics that help maintain coordination and memory by mopping up free radicals. According to a report in the *Journal of Neuroscience*, elderly laboratory rats that ate blueberries recovered some of the coordination normally lost with old age. Other antioxidant-rich foods include spinach and strawberries.[8]

## RESPECT FOR ELDERS

Protection and care by younger individuals can greatly improve an animal's chance of surviving into old age. But the elderly animal has to offer something in exchange for this care — something that enhances the fitness of the caregiver. In eusocial insects such as ants, there is often a division of labor according to age. Young Japanese queenless ants stay in the nest and breed, while older ants with resorbed ovaries work outside the nest, looking after the younger breeding females to which they are closely related. In mammalian species where "cultural wisdom" is often passed from one generation to another, elderly individuals have much to contribute to the well-being of their caregivers. The course and depth of underground water sources are remembered by older, more experienced elephants, memories that can literally save the herd from dehydration during drought. Conservationists are understandably concerned about the loss of elephantine cultural wisdom in Uganda, where 90 percent of the adults in Queen Elizabeth National Park were poached during Idi Amin's reign. Here young orphaned elephants have severe behavioral problems and are becoming a threat to local humans. They appear unable to find sufficient food for themselves and are running amok, killing farmers and raiding crops. Denied access to the knowledge and experience of their elders, they seem doomed to conflict with humans.[9]

Elderly nonbreeding lionesses with worn and missing teeth are not ostracized from their pride, but live out their old age — twenty years or more — supported by the hunting of younger females. Lionesses hunt using complex, flexible, cooperative strategies, which may benefit from the experience of older females. Chimpanzees also take care of their elderly, but usually a relative or an orphan without its own relatives does the caring. Wanaguma, a postmenopausal female at Mahale (who may have been in her early fifties), led a rather privileged and protected life under the watchful eye of her son. With failing eyesight caused by a cataract, a hunched back, and wobbly spider-like legs, she would have had difficulty surviving without his protection and attentive grooming. In chimpanzee society, old age brings what can only be described as respect. Although males normally are the targets of a great deal of aggression by other males, elderly males receive far fewer threats, and their own threats are tolerated without retaliation — even by younger, stronger males. Elderly chimpanzees are also the only members of society to be honored with more grooming than they give and to be allowed access to meat while others are rebuffed.[10]

Respect for the elderly is not universal among primates. It is not something that elderly baboons, for example, enjoy. Robert Sapolsky describes how it is to be old in a society of olive baboons in the

Massai Mara National Park: "Your average male baboon gets a lot of grief from the current generation of thugs, and this often leads to a particularly painful way of passing your golden years."[11]

It is interesting to consider which of our primate relatives we most closely resemble. In an industrialized society that tends to rely on young brains and new technology, we look for *new* answers to problems instead of seeking the wisdom our elders have accrued through years of experience.

## DEATH IN THE WILD

Although it is certainly true that the majority of animals die before reaching their full physiological life span, it cannot be said that *no* wild animal ever achieves it. Of the fifty-one chimpanzee deaths recorded at Gombe, six were apparently from old age. The rarity of seeing elderly wild animals nearing the end of their life span makes observations of them all the more poignant. Dian Fossey tells of an elderly African buffalo that regularly came near her camp in Rwanda: "From rump to withers his body was scarred like a road map, with countless healed wounds, possibly the result of encounters with poachers, traps, or other buffalo. The heavy boss on his head must have at one time been twice its size, but had been relentlessly chipped away over the years of his life. The remnants of the horns themselves gave evidence of decades of battles and remained only worn and shattered nubbins." After surviving so many battles, he finally lay down and died in a grassy hollow by her camp.[12]

The chance of dying a nonviolent death varies enormously for different species. Without man's hunting, more African elephants would probably die of starvation after their last teeth wear out. Poachers are the primary cause of death not only for African elephants, but for sea turtles, Siberian tigers, and mountain gorillas, to name a few. On the other hand, a rabbit is unlikely, even without human intervention, to live to a ripe old age, as a predator would catch any rabbit that showed signs of slowing down.

For many animals death is most commonly delivered by predators. But many violent deaths may be pain free owing to the stress-induced analgesia discussed in Chapter 11. One particular aspect of this phenomenon may be exploited by predators. When the sensitive tip of a horse's muzzle is contorted by twisting a piece of rope tightly around it, soporific endorphins (internal opioids) are released into the blood and the horse is drugged into placidity — sufficient so that it may undergo locally anesthetised operations such as castration. In light of this, it is interesting that when a pack of hunting dogs take down a zebra, there is often (although certainly not always) one animal hanging onto the zebra's muzzle with its teeth. Has the predator found an easy way to immobilize its prey through its own sedation reflex?

It is commonly believed that animals go off alone to die — as if, knowing their fate, they undertake some final ritualistic act. Cat owners grieve as their beloved sick pet disappears and they are left without a body over which to mourn. Dying gorillas seek to conceal themselves in the hollow holes of hagaenia trees. But "going off alone to die" is an unnecessarily complex interpretation of the behavior. Being seriously weakened makes an animal vulnerable to predation, hunters, or even other members of its own species waiting to take over its resources. It is not surprising to find seriously weakened animals attempting to hide and remain hidden until they have recovered, or died.

Legends of dying elephants separating from the group and making their way to special designated graveyards have persisted. Occasionally, on the plains of Africa, are found collections of many elephant bones, scattered here and there and tumbled together one on top of another. But no one has ever seen an elephant go off to a graveyard to die, despite many decades of study in the field. A more likely explanation for these bone piles is that old and sick elephants congregate in areas where vegetation is soft and easily digestible and where shade and water are nearby. Because many of these elephants then die, disproportionately more elephant bones appear in these areas. If this is so, the elephants do not go to the so-called graveyards in order to die, but rather in an effort to live. Another less generous, although perhaps more realistic, explanation is that elephant graveyards are the aftermath of a poacher's massacre.

Penguin graveyards have been reported in South America, although they have never taken on the mystical dimensions of elephant graveyards. South Georgia, a bleak island in the south Atlantic, is the main nesting ground for gentoo and king penguins. Aging and sick birds congregate around inland pools that are formed from melting snow, and the bottoms of these pools are covered with deep layers of penguin corpses.[13] In hot environments, there are often skeletons around watering holes, probably because dying animals try to stay near water.

## SUICIDE

Are humans the only creatures to commit suicide? Perhaps the most famous animal suicide myth is that of the Norway lemming, a fist-sized rodent of northern Europe. According to Scandinavian legend, every few years masses of lemmings descend from the birch woods into the lower pastures, destroying crops and driving on in their suicidal march to the lakes and seas below. A misleading wildlife documentary by Walt Disney in 1958 depicted lemmings jumping over cliffs to their deaths, but almost two decades later the film crew admitted that imported lemmings had been driven off the cliff to fit the script and the myth. In truth, lemming numbers periodically increase, and lemmings not being tolerant of each other, the overcrowding forces large numbers of youngsters to migrate en masse through rocky and difficult terrain. In such situations a sort of mass panic ensues, leading some lemmings to fall accidentally into steep ravines or occasionally into the sea.[14]

A Hawaiian story claims that toads periodically commit suicide by eating the flowers of the strychnine tree, which cause convulsions and death, but Ron Siegel found a more plausible explanation. Toads instinctively snap at floating objects falling past their view, and when the blossoms of this tree fall, the toads snap them up and accidentally poison themselves.[15]

In 1955 William Baze reported that an elephant he had captured and restrained committed suicide by walking round and round the tree trunk to which she was tied and when it was as tight as she could get it, "threw herself on to her knees and strangled herself."[16] I feel suicide is an erroneous interpretation of this event. Having had horses all my life, I am aware of the danger of tying a horse to an immovable object by an unbreakable halter. You can rapidly have a dead horse. Horses, like elephants, must be accustomed to their strength in comparison to their environment. There are few situations in the wild from which they cannot escape by sheer brawn. But where man-made materials such as ropes and chains are used, a horse can pull until it cuts off its own airway, or breaks its neck twisting and leaping. This is precisely how snares work in trapping foxes, rabbits, and hares — the animal pulls until it strangles itself.

Biologists only take a claim of suicide seriously when it is clearly adaptive. After a mother house spider has hatched her eggs, she donates her own body to feed her spiderlings, thus contributing to the propagation of her genes. When pea aphids are parasitized, they "commit suicide" by allowing themselves to be eaten by predatory ladybirds. Since the aphids are sterile, their future reproductive potential is zero. By dying with the parasites inside them, they prevent the parasites from emerging and infecting close relatives.[17]

Perhaps the behavior closest to what we might call passive suicide is when an animal gives up in the face of overwhelming odds. As we saw in the last chapter, certain individuals seem to lose the motivation to eat, clean themselves, and drink, slowly pining away.

## REACTION TO DEATH

The way animals react to a dying fellow can indicate much about their awareness of the health or vulnerability of others. A dying reindeer is simply left behind as the herd moves on, whereas a dying elephant or chimpanzee generates protective care, fear, or a display of grief.

Harvey Croze watched an elderly cow elephant in the final hours of her life. After lagging behind the family group for some time, she eventually lay down, dying. Initially her family tried to raise her, but soon they formed a protective circle around her and remained for the several hours it took her to die, and for several hours afterward. In most natural situations this protective behavior would be a great asset, providing time for an injured animal to recover. But when human intervention occurs, such protection can backfire. One research scientist darted a cow with anesthetic in order to fit a radio collar, but accidentally gave her too much sedative. When she collapsed, the other elephants closed up in a defensive phalanx, with the result that he was unable to administer any antidote. As he watched helplessly, the cow died.[18]

An extract from Cynthia Moss's *Elephant Memories* details the intensity with which elephants will try to help a close family member:

> Tina could go no farther. The blood pouring from her mouth was bright red and her sides were heaving for breath. The other elephants crowded around, reaching for her. Her knees started to buckle and she began to go down, but Teresia got on one side of her and Trista on the other and they both leaned in and held her up. Soon, however, she had no strength and she slipped beneath them and fell onto her side. More blood gushed from her mouth and with a shudder she died.
>
> Teresia and Trista became frantic and knelt down and tried to lift her up. They worked their tusks under her back and under her head. At one point they succeeded in lifting her up to a sitting position but her body flopped back down. Her family tried everything to rouse her, kicking and tusking her, and Tallulah even went off and collected a trunkful of grass and tried to stuff it into her mouth. Finally Teresia got around behind her again, knelt down, and worked her tusks in under her shoulder and then, straining with all her strength, she began to lift her. When she got to a standing position with the full weight of Tina's head and front quarters on her tusks there was a sharp cracking sound and Teresia dropped the carcass as her right tusk fell to the ground.[19]

This heart-wrenching account shows just how far elephants will go to help a dying relative — staying to help even with the hunters still close by. Elephants have also shown a willingness to protect other species. Heathcote Williams recounts a tale of a working elephant that refused to drop a large pillar into a hole while working on a building site. When angry workers looked down the hole, they found a dog asleep at the bottom. Scientists such as Joyce Poole and Cynthia Moss have seen elephants protect wounded humans; they appear to have an astute awareness of the vulnerability of others.[20]

## BURIALS AND FUNERALS

Stories of animal funerals and similar rites of passage are usually highly romanticized anthropomorphic interpretations of events. But elephants *do* attempt to cover their dead. Cynthia Moss described in detail how Tina's family reacted to her death:

> They stood around Tina's carcass, touching it gently with their trunks and feet. Because it was rocky and the ground was wet, there was no loose dirt; but they tried to dig into it with their feet and trunks and when they managed to get a little earth up they sprinkled it over the body. Trista, Tia and some of the others went off and broke branches from the surrounding low bushes and brought them back and placed them on the carcass. They remained very alert to the sounds around them and kept smelling to the west, but they would not leave Tina. By nightfall they had nearly buried her with branches and earth. Then they stood vigil over her for most of the night and only as dawn was approaching did they reluctantly begin to walk away, heading back toward the safety of the park. Teresia was the last to leave. The others had crossed to the ridge and stopped and rumbled gently. Teresia stood facing them with her back to her daughter. She reached behind her and gently felt the carcass with her hind foot repeatedly. The others rumbled again and very slowly, touching the tip of her trunk to her broken tusk, Teresia moved off to join them.[21]

An elephant will not pass the body of another elephant without covering it with twigs, branches, and earth or dust. After scientists and park officials had culled elephants in Uganda, they collected the ears and feet of the dead elephants in a shed, hoping to sell them later for making handbags and umbrella stands. One night, though, a group of elephants broke into the shed and *buried the ears and feet.*[22]

They will also bury the dead of other species. Katy Payne recounts an event in which a lion had pounced onto an elephant's shoulder: "In a single motion the elephant reached her trunk over the lion's body, grabbed him by the tail, ripped him off, and using the tail as a handle, slammed him onto the ground repeatedly until he was dead. The elephants then broke branches from some nearby bushes, and covered the dead lion with them."[23] And in India George Schaller saw an Asian elephant bury a dead buffalo put out to attract tigers.

European badgers also occasionally bury their dead, and two cases of badger "funerals" have been published in which badgers dug a hole, dragged a body to it, and heaped earth over it. Badgers will drag dead bodies for some distance and even cover roadkills with leaves. But a review of badger burials by Tim Roper concluded that they were probably storing the dead bodies as carrion.[24] Interestingly, sick badgers in captivity block their tunnel entrances with earth, and the majority of dead badgers in the wild are found blocked in their setts.

There is some evidence that gorillas may occasionally bury, or at least cover, their dead. The nineteenth-century anatomist Richard Owen reported that when a gorilla dies the others cover its corpse with a heap of leaves and loose earth collected and scraped up for the purpose, and John Berry found gorilla corpses covered in vegetation while working in Bwindi Impenetrable National Park, Uganda. Don Cousins wrote of captive gorillas discovering a dead crow in their compound in Woodland Park Zoo, Seattle: "After sniffing it, they scooped out a hole, flicked soil on the carcass, and buried it." Dian Fossey commented on how deceased gorillas in the wild seemed mysteriously to disappear; she assumed that the density and rapid growth of the vegetation must have concealed their bones.[25]

Many meat eaters bury carcasses for later consumption, but the covering of dead bodies in earth, leaves, and branches is not restricted to those that return to eat them. Elephants are, after all, vegetarian, and there have been no sightings of badgers eating the bodies of the badgers they bury. The fact that elephants cover wounds with earth and dust may provide a clue to why they cover the dead: both wounds and decaying bodies can spread infection. A coating of soil changes the pattern of decay quite markedly, preventing the early colonizers of the carcass (mainly calliphorid flies) from laying their eggs. A reduction in the number of corpse-colonizing flies could be a boon to those living nearby. Furthermore, a layer of dirt and vegetation is likely to reduce the scent of blood or decay that attracts scavengers and predators.

Hygienically disposing of the dead is not out of the question. Social insects are scrupulous about clearing corpses from their nests. Honey bees drag all their dead out of the nest to prevent the spread of infection, and when another animal becomes trapped inside the nest and cannot be dragged out (as when a mouse gets in and then dies of bee stings), the bees will coat the entire body in an antibacterial wax to prevent the spread of disease to the bees.

The time when we humans started to deliberately bury our dead is thought to coincide with a need for hygienic disposal of corpses. Historically, human burials became more common as groups became more settled. Early prehistoric humans simply moved on when their elders died. There were few, if any, deep burials before 2.5 million years ago, although it is impossible to determine whether earlier humans were covering their dead as do elephants, other primates, and badgers. But as human groups settled in protovillages, they started to bury their dead in graves within the settlement. Trevor Watkins, an archaeologist from the University of Edinburgh, explains: "Increasing frequencies of burials are associated with increasingly permanent settlements, where lack of deep burial would have been a serious hygiene risk."[26]

Many nonhuman species show apparent distress when their family or companions die. Huffman recorded this account of Mahale chimpanzees discovering the body of one of their group:

> At 11:00 the mid-morning calm was suddenly broken by a loud outburst of alarm calls. "Wrra, wrra!!" This unmistakable call is rarely heard except on occasions of great fear or alarm, for instance after hearing the deep raspy growl of a leopard prowling nearby in the forest. Half afraid ourselves of what lay ahead, we ran towards the commotion. The entire group was up in the trees peering down into the dense bush at something on the ground. The direction of the wind changed and the unmistakable smell of death told us all we needed to know . . . The chimpanzee group's response to the discovery of the body was much as I would expect that of humans to be if they came across the body of a fallen companion. First surprise and fear, then rage, followed by sadness. Several chimpanzees cautiously, almost reluctantly, approached for a look. A few individuals such as the adult male Fanana and the adult females Tootsie, Calliope and Nkombo spent several minutes each looking down at the body from the safety of the trees. The two young adults, Fanana and Linda, made day nests in the trees within three meters above the body. These two and later Cadmus, the five year old son of Calliope, softly vocalized, "hoo, hoo," at the body. They were concerned and seemed almost mournful. Fanana would not stare directly at the body for long periods of time, but preferred to turn his back and lay there quietly, glancing back on occasion. Others were more fearful and tried to steal a glance from a distance.[27]

Death is frightening in part because what killed the corpse might kill anyone else close by. Dead bodies are also a threat because a rotting carcass draws scavengers, predators, and flies, and may release pathogens into the water, air, or food supply. The fear that many animals appear to experience around death could therefore be an adaptation to an intense health risk. The archaeologist Mike Parker Pearson writes that in human societies "the dead are universally a source of fear, especially during the corpse's putrefaction."[28]

Many humans insist that other species do not understand death the way we do. Although we may have a superior intellectual understanding of death, the distress it causes may not arise because we are different from other animals, but because we *share* a biological fear and loathing of death — the antithesis of survival and the ultimate health hazard.

# 14

ꗥ ꗥ ꗥ ꗥ

# WHAT WE KNOW SO FAR

> The threat of parasites and disease should be considered as one
> of the important determinants of behavior.
> — Benjamin L. Hart, veterinary research scientist, 1991

I HOPE it is obvious from this brief foray into wild health that animals *do* medicate themselves against the ravages of microorganisms,
parasites, stress, and injury. The observations of many traditional
herbalists and amateur natural historians have proved valuable, but
it is also clear that much of the folklore concerning animal self-medication either has been based on incorrect interpretations of behavior
or is intended to be metaphorical. With the understanding that scientific investigation of this topic is very much a work in progress, let
us review what is known so far.

From the perspective of health maintenance, many seemingly familiar aspects of animal behavior take on fresh meaning. Foraging
habits, fear of strangers, territoriality, migration, disgust at fecal matter, and cannibalism all have roots in disease avoidance. And the examples of animal self-medication we have discussed here represent
only a tiny portion of those still to be uncovered. There has not been
room, for example, to explore fully the role of medicinal insects,
barks, mineral waters, sunlight, exercise, and plant fiber. Nor have we
discussed sufficiently what animals do to maintain their health during migration, hibernation, famine, and drought, or how they select
healthy mates to avoid disease. We have unearthed as many new and

exciting questions as we have answered. Even so, some important conclusions can be drawn.

One myth about animals that can be dispelled is that "they know unerringly which herbs will cure what ills."[1] There is little evidence that any animal acts as if it were a trained pharmacist, selecting a specific remedy for a specific ill. The thirty-four recorded species of hairy leaves used by chimpanzees to deal with the discomfort of intestinal parasites, the wide variety of aromatic greenery brought to the nest by different populations of starlings, and the range of skin rubs used by mammals all indicate that animals medicate themselves by flexible rules of thumb.

Flexibility is adaptive, because pathogens and poisons are always relative. Most microorganisms can be harmful or beneficial depending on their relative proportions. And compounds in nature's pharmacy can be poisonous, nutritious, or medicinal depending on dosage and circumstances. Health maintenance behavior must therefore be analyzed in context, and self-medication can only be understood from a holistic perspective; it is not easily reducible to specific mechanisms.

One characteristic of the many medicinal substances animals use is that they are multifunctional: they have numerous effects. As health threats come from all directions, from many sources, often at the same time, it is an advantage to have remedial strategies that take *broad action*. Eating earth, for example, can balance stomach acidity, line and protect the intestine, bind internal or dietary toxins, and provide essential minerals. It is not always apparent which particular benefit is paramount for any specific animal at any one time. Similarly, fur rubbing with pungent compounds has numerous potential benefits: repelling insects, soothing and healing tiny sores, stimulating superficial blood vessels, stimulating the immune system, and general arousal. This multifunctionality may be annoying to those who want to know which specific pathogen is being targeted, but it is a highly adaptive characteristic of self-medication. Unless we expect animals to have a magnificent grasp of microbiology, parasitology, and virology, we should not expect to find specific behavioral strategies for each ill. Multifunctional strategies will be far more beneficial to an animal that does not know what precisely ails it.

Animal self-medication is not confined by germ theory. In modern Western medicine, we treat disease by directly attacking the pathogen associated with the disease symptoms. A more holistic approach such as that seen in Oriental medicine assumes that the pathogen is not the direct cause, only a symptom of an imbalance — such as a disruption in physiological or psychological homeostasis. Indeed, we have seen how parasites and pathogens take hold during the stressful conditions of drought, famine, and overcrowding. From this perspective, attacking only the pathogens treats the symptoms of health disruption rather than the cause. Such distinctions are, of course, not available to animals. Not being acquainted with germ theory, animals are guided by how they *feel*, and this depends on the sum total of internal and external sensors rather than on the detection of particular pathogens. General rules of thumb that they can use to reinstate the sensation of "wellness" may therefore be more adaptive than carefully selected pathogen-targeted medicines.

Sometimes animals combine several broad-action strategies. Sick chimpanzees have been seen to take at least three actions to help themselves when they are suffering the discomfort of intestinal parasite infestation: they chew bitter-pith, which contains numerous multifunctional medicinal compounds; they swallow folded-up hairy leaves, which entrap worms and speed their expulsion from the gut; and they eat termite-mound soil, which has the physical and medicinal properties of clay along with microbial associations capable of secreting antibiotics.

Combinations of broad-action strategies may be one way in which animals reduce the all-too-familiar problem of pathogen resistance. This is something we humans have had to learn by trial and error. As the main parasites of our domestic animals now are resistant to our drugs, the latest approach is to apply a variety of deworming chemicals, each with slightly different actions. However, our approach is still primarily pathogen targeted. Similarly, we are finding that our hospitals are incubators of superbugs resistant to our powerful antibiotics; as a consequence, we are having to adopt more sustainable strategies based on administering a wider range of antimicrobials, and pay more respect to the natural avoidance strategy of careful hygiene.

Even with the limited data currently available on animal health maintenance, it is apparent that, in the wild, prevention is better than cure. Animals practice *constant vigilance* over their health, which is after all a matter of life or death. Selection will not have favored those individuals that waited until they were sick before paying attention to their health.[2]

What animals select to eat plays a vital role in preventive health care. Exposure to small amounts of toxins helps to stimulate the production of enzymes that will detoxify the same compound if it is eaten later; it can also enhance the immune system and generally stimulate metabolism. Early exposure to parasites and pathogens may prevent autoimmune diseases such as diabetes, multiple sclerosis, and inflammatory bowel disease from emerging later in life. But an important key to health maintenance is the way animals support gut microorganisms that competitively exclude other more dangerous pathogens. For example, the gorilla's natural diet of plant antibacterials contributes to a healthy cocktail of microorganisms that protects it from harmful intestinal bacteria. The gorilla's high-fiber diet also helps, as its fermentation creates the right level of acidity for the helpful microbes to survive. Eating volcanic clay may also be beneficial in this respect.

Throughout this book we have been concerned with the *intention* of animals that self-medicate. When highly intelligent primates such as chimpanzees suffer parasite infection, they show intention in that they deviate from their normal feeding behavior to seek out particular plants that help their condition. We do not know whether their intention is to relieve symptoms, to eradicate parasites, or to satisfy a new dietary craving created by parasite infestation. We do know that the results of this behavior are beneficial to the chimpanzee and detrimental to the parasites.

Evidence suggests that it is the removal of *discomfort* that motivates curative behavior. Young chicks with painful broken legs rapidly learn the benefits of eating analgesic food. Their legs are not mended, but their discomfort is eased. Their immediate somatic sensation — pain — is cured. Caterpillars seek out plants that reduce the symptoms of their internal parasites even though the parasites remain unharmed. The rats described in Chapter 5 ate clay when they felt sick, but — more important — they ate clay even when they were merely *conditioned* to feel sick. The sensation of well-being is a powerful influence on behavior.

We do not yet know how animals determine from a massive potential pharmacy what will make them feel better — although the same applies to diet selection generally. As with diet selection, we can expect the mechanisms of self-medication to vary between species and contexts. Caterpillars are obviously not using the same cognitive processes as chimpanzees. We have seen how sick rats observe what other rats are eating in order to find something that stops them from feeling sick, but they do not learn precisely what will help them — merely what is safe to try. In the end, they still must learn the consequences of what they eat for their *own* ills.[3]

Primates, with their greater cognitive skills, are capable of learning more directly about nature's pharmacy. Spider monkeys released from captivity into Barro Colorado Island quickly learned the value of citrus fruits as skin rubs, even though they had not come across them before. Similarly, the charcoal-eating red colobus monkeys of Zanzibar are apparently passing the habit from mother to offspring.[4]

We can obtain some clues about how intelligent animals learn about nature's pharmacy by asking how traditional herbalists narrow down possible medicines. The Mixe Indians of Oaxaca, Mexico, use plants with astringent properties for diarrhea and dysentery, whereas bitter-aromatic plants are employed for gastrointestinal cramps and pain.[5] Animals seem to utilize general guidelines in this way too. Most of the herbs birds use to enhance the health of their chicks, and the skin rubs mammals and birds use, are aromatic-astringent. The plants apes, geese, and bears use to scour their intestinal parasites are all rough in texture. Through sampling, a bird or mammal could readily come to associate certain smells, tastes, and textures (rather than specific plants) with particular physical sensations.

One aspect of health care that clearly distinguishes animal medicine from our own is that there are few animal "doctors" administering medicines to others, and no documented cases of one animal teaching another how to use medicines (although there are examples of learning by observing others). If self-medication is about reestablishing a feeling of well-being, there is little of relevance that one individual can pass on to another.

What we have uncovered so far are not animal herbalists but animals constantly striving to reinstate a feeling of physiological and psychological well-being. Natural selection will have dispassionately weeded out ineffective strategies.

Wild health is a dynamic interaction of physiology, behavior, and the environment, and it is the constant attention to well-being we see in the wild that is missing in industrial human society, as well as in our management of animals in our care.

# PART III

# LESSONS WE MIGHT LEARN

❉ ❉ ❉ ❉

# 15

# ANIMALS IN OUR CARE

Disease is the biggest animal welfare issue today.
— Robin Pellew, director of the Animal Health Trust, 1992

ANIMALS IN CAPTIVITY, unable to range freely over their natural habitat, rely on humans for their health care. Sadly, we often fail to do an adequate job, but applying what we have learned about wild health can greatly improve this situation.

## ZOO ANIMALS

Captive-born wild animals often live longer than those in the wild. Longevity, though, is not necessarily a valid indicator of health: it is simply the removal of fatal events. Captivity actually causes much illness in wild species. By far the most common effect is some kind of psychological disturbance, manifested in abnormal behavior such as stereotypical movements, self-abuse, or at worst a catatonic state of complete sensory shut-down. However, captivity also causes physical ill health. To cite a few examples from a vast selection, nearly half of zoo gorillas die of cardiovascular disease, and many are infertile or suffer eating disorders. Captive elephants sustain foot problems and arthritis, and often lack the urge to mate. Captive giraffes suffer arthritis and excessive hoof growth. And the hemolytic anemia that kills 75 percent of its captive black rhinoceros victims is not seen at all in wild rhinoceroses.[1]

It is not surprising that health is compromised by captivity. Animals are denied access to the food, water, soils, microorganisms, climate, weather, grooming, and wound tending to which they are adapted. At the same time, from close association with other animals and humans, they are exposed to new pathogens to which they have no resistance. Captive gorillas have to be protected from humans by glass barriers if they are to avoid serial infections. Captive animals cannot move away from pathogen hot spots and cannot avoid other animals that may be violent — and they are often isolated from health-enhancing social interactions.

Diet alone accounts for much ill health in zoos. We know little about the nutrient requirements of most wildlife species, and only in 1994 did the American Zoo and Aquarium Association set up a group to promote nutritional science in animal husbandry. Currently most zoo diets are based on tradition. The Asian elephant, for example, is routinely fed a diet extrapolated from the nutritional needs of a large horse. Standard rodent diets are fed to all species of rodent, from the carnivorous grasshopper mouse to the herbivorous capybara.[2] There are no standard guidelines for the zoo diets of colobine primates, and ordinarily they get much less fiber than they would in the wild. Captive woolly monkeys commonly die of kidney and liver failure caused by an inadequate diet. Flying foxes die of congestive heart failure due to insufficient vitamin E; fruit-eating birds accumulate hepatic iron stores; and carnivores and many primates suffer obesity and persistent diarrhea, along with tooth and urinary problems. Captive reptiles frequently suffer rickets and osteoporosis because of inadequate calcium and phosphorus. Often they stop eating altogether and have to be force-fed.[3]

As we have seen, nutritional and nonnutritional dietary needs change with health, reproductive status, age, and social environment. An animal frequently "knows best" what it needs, through adaptation or learning. Under experimental conditions, mice have shown that they are better than their human caretakers at mixing exactly the right proportions of dietary constituents to optimize health.[4]

As diet is still so poorly understood, it is little wonder that any self-medication needs are almost totally ignored. Those crucial "medi-

A woolly monkey
in the Apenheul
Zoo, Netherlands,
inspects a plant
in her enclosure.
*Warner Jens*

cines" that we have seen contribute so much to preventive and curative health are not yet considered part of a captive animal's requirements: bioactive substances are needed as skin rubs by bears, coatis,
monkeys, and birds; grasses and rough leaves are needed as scours
and emetics by big cats, wolves, bears, geese, and chimpanzees. The
provision of such plant matter might greatly enhance the health of
these species.

In a pioneering move, Holland's Apenheul Zoo introduced a range
of medicinal plants into its primate enclosures. Keepers noticed that
after fights many of the monkeys started to feed on valerian (*Valer-
iana officinalis*) — an herb used by herbalists as a sedative and as a
hypotensive to lower blood pressure. In captivity these particular
monkeys have long suffered from dietary hypertension that leads to
kidney and liver failure. The Duke University Primate Center provided a similar but smaller sample of herbs to its captive apes, but

they showed a disappointing lack of interest.[5] The difficulty in both situations is that what will grow in the enclosure is not the same as what grows in the primates' natural habitat. Furthermore, the plant selections are based on what humans think the animals might need.

Many seemingly strange behaviors of captive wild animals could be attempts at self-medication. Plant eaters commonly consume clay in the wild, and apparently abnormal digging behavior by many captive species could be indications of this need. Gorillas injure themselves trying to dig into concrete floors, and traveling circus elephants crunch assorted rocks they find on tour.[6] Both species mine rock for self-medication in the wild.

In addition, access to appropriate levels of shade and natural sunlight are essential to health. In early zoos, locking wild animals out of the light caused many health problems. Captive primates suffered terrible bone deformities and cannibalized their litters, until sunlight was found to be vital in providing enough vitamin D to keep them healthy. Camels too sunbathe in the wild, and get sick if deprived of light. Chameleons reared indoors usually suffer ill health if not provided with artificial UV-B irradiation; without it, they and their eggs die from vitamin deficiency. Even bacteria need light to repair damaged DNA.[7]

In captivity it is difficult for an animal to get adequate exercise. A free-ranging tiger that can wander over 50 kilometers every night, year in, year out, in captivity may have no more than 100 meters available if it is lucky. An elephant that would normally travel hundreds of kilometers searching for the plants it needs to fulfill its health requirements often finds itself confined in a concrete house a few tens of meters across. Unsurprisingly, free-ranging wild animals have greater heart strength than confined animals.[8]

Captivity also restricts social interaction. Being confined with bored, psychologically disturbed fellows can lead to violence and self-abuse. Unable to escape, a bullied animal can be repeatedly wounded unless removed to another enclosure. Even the simple act of observation can impair the health of certain species: gorillas threaten one another with direct eye contact, and a continuous stream of staring people can greatly disturb captive gorillas.

Zoo design has vastly improved since the early days of small concrete-floored cages. Most zoos today attempt to enrich the animal's environment. Food may be hidden and varied to make foraging more interesting, pools are provided for cool dips, as are dust for dust bathing and hiding places in which to escape those prying eyes. Unfortunately, though, animal health is not a priority in all zoos. The bear parks of Japan are a case in point. In Okuhida Bear Park in Kamitakara-Mura, Honshu, crowded groups of bears (normally solitary animals) are displayed in bare concrete pits, where they are forced to beg for food from visitors. When the violence between them gets too ferocious, keepers spray a water jet into the pit. Younger bears wait to enter the pits, locked below ground level in cramped cages so small they cannot fully stand. The visitors, having been entertained by the fighting and begging, can buy tins of bear meat in the park shop on the way out.[9]

Although captivity can be detrimental to health, there are times when the confines and protection of humans are actively sought. Elephants of the national parks in Africa and India, for example, run back inside their boundaries when hunted by poachers. Similarly, kudu attempt to jump the high fences of national parks to get *inside*, where grazing is better managed. However, these protected areas are very different from display enclosures.

Observations of wild health can help us better protect endangered species in the wild. Habitat destruction is the main threat to wildlife health. The elephants of South Africa have lost 20 percent of their range in just ten years and remain isolated and restricted in a mere fragment of the wild lands their ancestors once roamed. This loss of habitat reduces the opportunities for health maintenance, although some species may be able to take compensatory action. On the slopes of Mount Kenya, African buffaloes that normally inhabit the lowlands have been forced up into the mountains by human encroachment. They counteract the anemia they experience at high altitudes by mining subsoils rich in iron.[10]

As groups of animals inhabit ever smaller islands of natural habitat, signs of inbreeding start to appear in the form of unusual disease and ill health. In Brazil, tree-dwelling black lion tamarins, isolated in a disjointed 5 percent of a once vast forest of São Paulo, have to be transported by conservationists across the treeless plains to find new mates. However, habitat destruction not only deprives endangered species of food, shelter, and genetically novel mates, it leaves them bereft of plants or soils essential to their health. Because medicinal substances may be used only rarely, conservationists can easily fail to recognize their importance. The more we know about how animals keep themselves well, the better we will be able to ensure that protected areas provide everything that is needed.

As human activities reduce wildlife populations, it is tempting to boost falling numbers by reintroducing captive-bred animals into the wild. This procedure has the added bonus of contributing new genetic material. But if, as we have seen, nutritional wisdom and the ability to self-medicate are largely learned skills, not only will captive-bred animals be considered strangers by their wild fellows, but they will need to learn how to gain sufficient nutrients in an unfamiliar and ever-changing environment, avoid poisoning themselves, avoid pathogen hot spots, and deal with parasite infestations. Furthermore, they will not have the benefit of any early exposure to local toxins and pathogens. This explains why, in decades of international conservation, there have been only a handful of successful reintroductions.

## OUR COMPANION ANIMALS

Domestic dogs and cats have been selectively bred for characteristics attractive to humans — temperament, appearance, color — not for their health maintenance skills. One might not expect them to show much residual ability to actively maintain their own health. Sharing our lifestyles and eating the food we provide, our companion animals share many of the diseases of industrialized society. Dogs are increasingly diagnosed with allergies, obesity, dementia, and neuroses, horses with allergies to their basic diet of grass and hay. Many of these ills appear to respond well to dietary regimes based on those of their wild relatives.[11]

Our domestic cats and dogs still eat grass like their wild ancestors. It seems rough grass can help cats regurgitate fur balls or scour the gut of intestinal parasites. Denied suitable grass, cats will attempt to chew on houseplants. Some behavioral problems of cats, such as wool chewing in Siamese cats, may have its roots in ancestral self-

A domestic cat chewing grass (or house plants) may be trying to self-medicate. *Your Cat* magazine

medication, as may many behaviors of dogs. Dogs, domesticated from wolves back in prehistory, are probably attempting self-medication when eating charcoal, chewing grass, eating or licking sand, soil, and clay, drinking urine, and eating strange objects such as feces, rocks, or socks.

Wendy Volhard has been breeding Landseer Newfoundland dogs for more than twenty years and has, rather unusually, kept them in family units of mother, father, and pups. She has observed fathers teaching the pups what plants to eat: the father passes the plant through his mouth, leaving a trail of saliva for the pups to follow and sample. One plant the dogs commonly eat is goldenrod (*Solidago* genus), which contains saponins, diterpenes, phenolic glucosides, and tannins. Herbalists use its anti-inflammatory, antifungal, and antiseptic properties. All breeds of dog are not the same in this respect.

She has never seen her standard wirehaired dachshunds eat plants of any description (not even grass), but they eat dirt every day. These dogs scavenge carrion from fox kills in the woods, and the daily dirt could help prevent the buildup of internal parasites, since the dogs never need to be dewormed. Scientific studies of anecdotal observations such as these are sorely needed.

Captive parrots and macaws are famed for their self-abuse — a sure sign of psychological illness. As they would naturally be part of a huge flock of perhaps a hundred other birds, it is not hard to see why solitary confinement might make them unhappy. They also suffer a range of illnesses from pathogens successfully carried by free-ranging birds, and up to 30 percent of captive budgies, parrots, cockatoos, and cockatiels die of cancer. In the wild, these birds regularly consume clay as a means of detoxifying their food, yet in captivity clay is often not provided, even though the birds may be fed a diet rich in secondary plant compounds. The accumulation of toxins may contribute to ill health.[12]

## MEAT MACHINES

Nearly all the strategies for wild health are denied to intensively reared factory farm animals. Stocking densities are *much* higher than in natural populations, so parasites and disease spread easily and rapidly. Confinement means it is hard for animals to avoid pathogens and parasites or enjoy the health-enhancing effects of exercise and sunlight. Animals are also prevented from changing their diet as conditions change; they cannot select the right proportions of nutrients, or essential nonnutrients, to balance intestinal microorganisms, acidity, or immune function.

Farm animals have been selectively bred for "productivity" — the ability to turn food into meat or animal products for human consumption. In modern industrial farming, as long as the animal survives long enough to get to the abattoir, its health along the way is one of the least significant elements in its breeding.

The diet we provide confined farm animals bears little resemblance to their natural diet. We feed animal waste to our vegetarian livestock, adapted to browse or graze a range of plants. In North

America, chicken excrement (from intensive hen houses) is fed directly to confined cattle to provide them with protein and uric acid. Occasionally a dead bird gets mixed in, and an outbreak of botulism ensues. As we saw in Chapter 3, free-ranging cattle can successfully avoid botulism even when seeking minerals from bones. In 1999 the French government admitted illegally feeding human sewage sludge to its cattle.[13] And in Chapter 6, we saw how cattle that are fed the ground-up brains and spinal cords of other cattle developed mad cow disease. They have passed it on to humans in the form of new variant Creutzfeldt-Jakob disease (nv CJD). Had we learned from the adaptive strategies of wild animals, we would have known that herbivorous ruminants naturally avoid eating animal meat and never engage in cannibalism. The health maintenance strategies of wild animals have been tried and tested over millennia; they are not mere romantic folklore.

Chicken farmers too could learn valuable lessons from wild health. Red jungle fowl live in the forest in small groups of fewer than twenty birds, with one cockerel controlling and protecting a number of hens. They scratch around on the forest floor, finding insects, worms, and fresh greenery to eat. They dust-bathe and sun their feathers to keep them clean and healthy, and when it rains, they preen themselves all over. At night they roost up in the trees. Their claws are specially adapted to clasp thin branches even while they sleep, keeping them safe from predators patrolling the forest floor.

Broiler chickens are a long way from their wild ancestors. The poultry industry has continuously selected birds that turn food into meat as quickly as possible. In the early 1980s, it took eighty-four days for a broiler chicken to grow to weight. The time today is about half as long. This selective breeding means that muscle is being laid down *before* the circulation and heart have developed sufficiently to support the huge muscle load. As a consequence, the birds suffer circulatory problems and heart failure. On top of this, their bones are not strong enough to support their extra body weight, and lame birds die of thirst or starvation because they are unable to reach the automated food and water supplies. Up to 80 percent of broilers suffer broken bones or other skeletal defects at any one time, and seventeen thousand birds die of heart failure each day in the United Kingdom. A computer program selects their food according to a "least-cost formulation," which calculates the cheapest ingredients available at a specific time that will provide basic nutrients. These might include ground-up feathers, banana skins, coconut husks, or even chicken carcasses.[14]

To enhance their growth rate still further, feed is supplemented with antibiotics — although this practice is being phased out owing to concerns over adverse effects on human health. The birds are kept indoors in dim lighting lest they get "excited" and attack one another. They trample on their dead companions, blister their feet in the acidity of their own excrement, and damage their lungs in an atmosphere of ammonia fumes, dust, and bacteria.

Not surprisingly, birds in poor health present a health threat to the humans who eat them. The United Kingdom's Central Veterinary Laboratory found that almost half of broiler flocks are colonized with *Campylobacter*, which in humans can cause nausea, headache, backache, fever, abdominal pain, diarrhea, and in some cases arthritis and neurological problems.[15] *Campylobacter* food poisonings are increasing, and although the poultry industry would like to blame the rise on inappropriate cooking or contamination at the abattoir, research shows that the meat becomes contaminated with such organisms in the production stage — on the farm.[16] Although the presence of a pathogen is not the same as the presence of the disease, we have already seen many examples where human health is definitively linked to the health of our farm animals, and specific strains of *Campylobacter* have been followed from chicken to consumer.[17]

As we saw in Chapter 7, even these pathetic birds are able to self-medicate the pain of their lameness when given the opportunity. And this is not the only form of self-medication of which they are capable. During hot weather the birds suffer from heat stress in the poorly ventilated sheds, and many of them die. On the morning of June 20, 2000, after an uncommonly hot night, twelve hundred broiler chickens were found dead in their automated production unit at Kentford, in my own home county of Suffolk.

Research on animal self-medication could have helped. It has long been known that supplementing chicken feed with vitamin C (ascorbic acid) helps chickens cope better with heat stress; but giving extra vitamin C to nonstressed birds causes other health problems, so producers have difficulty knowing when, and by how much, to supplement the feed. Michael Forbes and his colleagues at Leeds University found a clever way to solve the problem. They realized that the birds might be able to self-medicate if they had some way of *detecting* the tasteless, colorless, odorless vitamin C. Birds have acute color vision and readily learn color associations, so the scientists simply colored food spliced with vitamin C and not the rest. Within *three days* the birds had recognized the positive effects of the colored food and learned to self-medicate with it, as and when necessary. Forbes's team believes that vitamin C works by reducing the production of the stress hormone corticosterone and thereby reducing other symptoms of chronic stress. They point out that self-medication with vitamin C could be applied to other forms of stress, such as parasite infection, high humidity, and high production rates.[18]

## FREE-RANGE SELF-HELP

Cattle too will medicate themselves, given the opportunity. Bill Roundy remembers how, a generation ago, he and other ranchers in Utah learned a valuable lesson by observing their sick cattle. Whenever a cow suffered prolonged scouring (diarrhea) and went off her food, the ranchers turned her out to fend for herself. But the sick cow frequently returned after a few days, fully recovered and ready to feed with the rest of the herd. Sickly cattle would travel many kilometers to clay banks and feed on the clay until their health returned, so the ranchers took to transporting clay to their stock as a preventive measure — a practice still followed today.

In the mountains of Venezuela, where cattle range freely over vast areas, they dig into ancient subsoils to access the clay they need. Research shows that clay benefits cattle by absorbing endotoxins and viruses such as bovine rotavirus and coronavirus. Some barn-reared cattle get clay fortuitously when it is used as a binding agent in cattle pellets. Others (particularly in North America) are routinely supplemented with bentonite clay, because growth improvements are a bonus of reducing gastrointestinal problems. Preweaned calves, though, have no access to clay, and in 1992 over half of heifer dairy calves in the United States died of scours. In the United Kingdom, the benefits of clay have not been generally accepted and an estimated 170,000 dairy calves die each year of diarrhea associated with bacterial infections. One explanation offered by the Food Standards Agency is that food and medicine are dealt with by different departments within the organization, and clay is considered neither one nor the other. It seems our cultural separation of food and medicine can have major implications for the health of animals in our care.[19]

Sheep are rarely credited with much know-how, but when it comes to health maintenance, they are gaining quite a respectable reputation. They avoid forage contaminated with feces, thereby reducing their exposure to parasite larvae. When infested with nematode parasites, they graze on particular plants that significantly *reduce* the infestation. In New Zealand, for example, parasitized lambs selected the bitter and astringent puna chicory and thereby reduced their parasite load. Tannin-rich plants such as this one are commonly selected in moderate amounts by free-ranging animals, and scientists from Australia and New Zealand have found that certain types of forage, such as *Hedysarum coronarium, Lotus cornicularus,* and *L. pedunculatus,* which contain more useful condensed tannins, can increase lactation, wool growth, and live-weight gain in sheep, apparently by reducing the detrimental effects of internal parasites. Cattle can help themselves reduce bloat on these forages, too.[20]

Herbalists have noticed that lambs eat quantities of earth when suffering worm infestations. Research at the University of New England, Australia, has shown that bentonite clay in the diet increases the flow of both dietary and microbial protein to the intestines and has a beneficial effect on wool production. From what is known about geophagy, it might be prudent to provide safe, fine-textured clay for all hoofed livestock to eat whenever they desire. Sheep also retain an ability to balance their gut microflora, and adult sheep can shed an infection of the lethal bacterium *E. coli* 0157 by being allowed outside onto suitable browse (see Chapter 6). It may be that sheep retain more of their natural health maintenance skills because they have undergone less domestication than other farm animals.

Further studies of self-medication in livestock species are essential to the rapidly developing field of ethnoveterinary medicine, in which local home-grown remedies are used to help resolve animal health problems.

Many intensively reared farm animals are depressed. This statement is not anthropomorphic: distress and depression are recognized pathologies of livestock animals, indicated by a marked reduction in general activity and a lack of reactivity.[21] Under intensive farming conditions, pigs sit for long periods on their haunches, heads drooping and eyes half closed — a sign of deep depression. Unfortunately, nearly all the coping strategies used by animals to reduce stress in the wild are restricted in captivity. But allowing our farm animals to follow their own health-enhancing strategies is not purely an animal welfare issue. Contented animals require fewer drugs, which not only keeps costs down but reduces any adverse effects on human health of eating meat laced with drug residues.

Some argue that it is expensive to provide the conditions necessary to keep farm animals content and healthy. Because human health is directly reliant on the health of the food we eat, however, we risk paying for cheap food with our health. By gaining a better understanding of the ways in which farm species naturally manage their own health, we might reduce sickness and the surfeit of medication they require, increase the quality of produce, and at the same time improve animal welfare. By working *with* nature rather than against it, we could save time, effort, money — and our own health, as well as the health of our animals.

# 16

Ⅺ Ⅺ Ⅺ Ⅺ

# HEALTHY INTENTIONS

> The only way to keep your health is to eat what you don't want,
> drink what you don't like, and do what you'd rather not.
> — Mark Twain, 1897

ONE HUNDRED YEARS AGO, the leading causes of death in the industrial world were infectious diseases such as tuberculosis, influenza, and pneumonia. Since then, the emergence of antibiotics, vaccines, and public health controls has reduced the impact of infectious disease. Today the top killers are noninfectious illnesses related essentially to lifestyle (diet, smoking, and lack of exercise). The major causes of death in the United States in 1997 were heart disease, cancer (of the breast, colon, and lung), and stroke. Chronic health problems such as obesity, noninsulin-dependent diabetes, and osteoporosis, which are not necessarily lethal but nonetheless debilitating, are steadily increasing, and our psychological health appears to be deteriorating at an alarming rate. In the United Kingdom, suicides by young men have increased by 176 percent since 1985, and according to the World Health Organization, depression currently disables 20 percent of the global population. Economic and technical progress is no assurance of good health.[1]

Humans are qualitatively different from other animals because we manipulate the flow of energy and resources through the ecosystem to our advantage, and consequently to the detriment of other organisms. That is why we compete so successfully with other species. But with this success come some inherent failings, particularly in terms of our health.

According to physician Boyd Eaton and his anthropologist colleagues, despite all our technological wizardry and intellectual advances, modern humans are seriously malnourished. The human body evolved to eat a very different diet from that which most of us consume today. Before the advent of agriculture, about ten thousand years ago, people were hunter-gatherers, living on fruit, vegetables, and lean meat — the foods varying with the seasons and climate, and all obtained from local sources. Our ancestors rarely, if ever, ate grains or drank the milk of other animals. Although ten thousand years seems a long time ago, 99.99 percent of our genetic material was already formed. Thus we are not well adapted to an agriculturally based diet of cereals and dairy products.

Our genes evolved to work in an environment of a hunter-gathering diet. Although adaptation has continued, at least one hundred thousand generations of people were hunter-gatherers; only five hundred generations have depended on agriculture, only ten generations have lived since the onset of the industrial age, and only *two* generations have grown up with highly processed fast foods. There has simply not been time for our bodies to adapt to such a dramatic change.[2] Physicians Randolph Nesse and George Williams write: "Our bodies were designed over the course of millions of years for lives spent in small groups hunting and gathering on the plains of Africa. Natural selection has not had time to revise our bodies for coping with fatty diets, automobiles, drugs, artificial lights, and central heating. From this mismatch between our design and our environment arises much, perhaps most, preventable modern disease."[3]

Do we really want to eat like prehistoric humans? Surely "cavemen" were not healthy. Surely life was hard and short. Apparently not. Archaeological evidence indicates that these hunter-gatherer ancestors were robust, strong, and lean, with no sign of osteoporosis or arthritis — even at older ages.

Paleolithic humans ate a diet similar to that of wild chimpanzees and gorillas: fresh raw fruit, nuts, seeds, vegetation, fresh untreated water, insects, and wild-game meat low in saturated fats. Much of their food was hard and bitter. Most important, like chimpanzees and gorillas, prehistoric humans ate a wide variety of plants — an estimated one hundred to three hundred different types in one year. Nowadays, even health-conscious Westerners seldom consume more than twenty to thirty different species of plants. A broad range of plants provides not only essential vitamins and minerals but also valuable secondary plant compounds integral to preventive and curative medicine.[4]

The early human diet is estimated to have included more than 100 grams of fiber a day. Today the recommended level of 30 grams is rarely achieved by most of us. Humans and lowland gorillas share similar digestive tracts — in particular, the colon — but while gorillas derive up to 60 percent of their total energy from fiber fermentation in the colon, modern humans get only about 4 percent. When gorillas are brought into captivity and fed on lower-fiber diets containing meat and eggs, they suffer from many common human disorders: cardiovascular disease, ulcerative colitis, and high cholesterol levels. Their natural diet, rich in antioxidants and fiber, apparently prevents these diseases in the wild, suggesting that such a diet may have serious implications for our own health.[5]

Contemporary hunter-gatherer societies still eat in the traditional way and, like our prehistoric ancestors, have far less cancer, heart disease, diabetes, and osteoporosis than those of us who forage in supermarkets. "Sickness is much more prevalent, pervasive, and diverse among agriculturists than among hunter-gatherers," explains anthropologist Michael Logan,[6] and it is not because the hunter-gatherers die before these ills can develop. A decline in health is seen with the transition to agriculture. As native tribes of what is now the south central United States abandoned their hunter-gathering lifestyle some fifteen hundred years ago, their worsening health was indelibly recorded in their skeletal remains. Although more of them *survived* the famines than did their ancestors, they were not as healthy. They were less robust and showed signs of deficiencies of vitamin B, iron, and protein. Agriculture overcomes the vagaries of seasonal food supply, but at a price. And as contemporary hunter-gatherers change to an industrialized diet high in fats and sugar and low in fiber, they too develop the diseases of the industrial world.[7]

Agriculturists select and domesticate plants for ease of cultivation and palatability. Over time they have chosen plants with fewer bitter-tasting or astringent secondary compounds, and these plants are inevitably more susceptible to disease. Modern crops, therefore, need more chemical intervention than wild plants, which retain their own defensive pesticides. Consuming modern crops is consequently very different from consuming wild plants, and when we eat the meat of domesticated animals fed on these domesticated plants, our total intake of beneficial plant compounds is far lower than if we had eaten wild game.

Furthermore, agricultural biodiversity is shrinking as fewer species and varieties are made available for cultivation. Today 75 percent of the global food supply comes from a mere twelve crop species.[8] Not only are we losing species diversity but we are losing varieties within those species. The demise of dietary diversity is exacerbated by modern processing, in which artificial chemicals instead of herbs are used to preserve, enhance the taste of, and add color or other properties to food. Our industrial diet is greatly weakened thereby in both nutritional and medicinal attributes, providing us with only the bare essentials of energy and protein.

Not all agricultural societies have taken the same road. Many traditional agriculturists maintain the diversity of their diet by eating a variety of herbs and other plant compounds along with meat and grains. The Huasa people of northern Nigeria, for example, traditionally include up to twenty wild medicinal plants in their grain-based soups, and peoples who have become heavily reliant on animal products have found ways of countering the negative effects of such a diet. When animal fat is metabolized in the body, it produces damaging free radicals that contribute to cardiovascular disease, cancer, and aging.

While the Masai of Africa eat meat and drink blood, milk, and animal fat as their *only* sources of protein (animal fat makes up 60 percent of their energy intake), they suffer less heart trouble than Westerners. One reason is that they *always* combine their animal products with strong bitter, antioxidant herbs — up to twenty-eight additives in each meat-based soup, and twelve substances added to milk! In other words, the Masai have balanced the intake of oxidizing and antioxidizing compounds. According to Timothy Johns, it is not the high intake of animal fat, or the low intake of antioxidants, that causes so many health problems in industrial countries; it is the lack of *balance* between the two.[9]

## TOO MUCH OF A GOOD THING

Flavonoids are among the many plant compounds that have protective properties, and food technologists keen to tap into the lucrative health-food market are promoting genetically engineered vegetables with high levels of flavonoids as "functional foods": tomatoes with ten times more lycopene than ordinary tomatoes (lycopene reduces the risk of cancer and heart attack); and superbroccoli with ten times as much sulforaphane, a compound that stimulates production of enzymes which destroy cancer-causing substances in the gut.

Yet more of one single compound is not always better. We have seen that the difference between a toxin and a medicine is often dose related, and that small amounts of certain compounds can be protective while larger amounts are damaging. This is true of something as ordinary as salt or sugar. The food industry adds them to our already heavily processed diet to enhance our pleasure when consuming its products. Unfortunately, too much salt and sugar are common causes of high blood pressure and diabetes, respectively. We *can* have too much of a good thing.

We do not need to engineer health-enhancing food. We simply need to eat sufficient quantities and varieties of fresh organic fruit and vegetables that still retain their own protective plant compounds. In a review of the impact of diet on cancer, Donald Hensrud and Douglas Heimburger of the Mayo Clinic conclude, "The strongest and most consistent protective factors against gastrointestinal cancer are vegetables and fruit." They add that the best effects are seen from consumption of *whole* plant food rather than extracted isolated ingredients.[10] Foods and herbs with the highest anticancer activity include garlic, soybeans, cabbage, ginger, licorice root, and the umbelliferous vegetables such as broccoli and cabbage. Citrus fruits contain a host of valuable plant compounds, along with vitamin C, folic acid, potassium, and soluble fiber. Clinical trials have not yet demonstrated the same protective effects from taking supplements as from eating real food.[11] Our observations of wild health all point in the same direction — that we should eat more fruits and vegetables, not only for nutrients and energy but for essential health-enhancing nonnutrients too.

## WILD COUCH POTATOES

Lest we start to imagine that wild animals have an inner wisdom that leads them always to make the best decisions for their health, we should take heed of the many situations in which they, like us, have eagerly succumbed to an easy life. The health strategies we see in this book *are only effective in the context in which the animal evolved,* that is, where food is often scarce, seasonal, and highly varied, and where nutrients are hard to come by. A baboon, for example, can spend up to 40 percent of each day searching and foraging for the nutrients and energy it needs. As a result, the natural baboon diet is high in fiber and low in sugar, salt, fat, and cholesterol.

Given the opportunity to become couch potatoes, baboons jump at it. Robert Sapolsky has for many years studied a group of baboons in the Masai Mara National Reserve on the Serengeti Plains of Kenya. He has seen how they scrape and forage — picking out a little sweetness here, a little oil there. But as it has attracted more tourists, the park has had to dispose of larger amounts of waste from hotels. Within a few years of the creation of the first waste dump, the local baboons learned that they did not need to forage all day. They could lie in bed till the waste truck arrived, binge on high-sugar, high-fat, high-protein junk-food leftovers, then relax all afternoon. Over the years, Sapolsky has watched his junk-food monkeys (as he calls them) change. They grow faster as youngsters, reach puberty earlier, and weigh more. Their cholesterol and insulin levels have gone through the roof, setting them up for chronic heart disease and adult-onset diabetes. Not long ago, the whole troop was nearly wiped out by bovine tuberculosis from contaminated meat thrown out by a local hotel. Although baboons normally avoid eating cow meat, they did not in this situation.[12]

A similar scene occurs in North America, where wild bears hang around the waste dumps of forestry workers and the car parks of picnicking tourists in such places as Yosemite National Park. They become obese — often twice their normal weight — poisoned on plastic and metallic waste, confused about hibernation, and violent as they fight over access to the high-energy junk food.

Animals (including most of us) have no long-term health aims only short-term goals. Sweet is good because it means energy, so "lots of sweet" means "lots of good" means "lots of energy." Fat is good because it means lots of energy for storage, therefore "lots of fat" means "lots of energy," even in times of famine. In other words, our mammalian sensory system and physiological feedback systems help us stay well only if our choice is limited to real food — vegetables, fruits, game — in the small but frequent amounts available to our foraging ancestors. There has been no selective pressure to stay slim or healthy by conscious self-denial.

We should be wary of foods that exploit distortions in our innate nutritional reward system. Chocolate contains lots of fat, lots of sugar, and a slightly bitter stimulant that locks into our brain pleasure centers. No wonder we love it! We crave the food sensations that in the past were needed to keep us motivated to forage for hours every day. When energy was scarce and food was natural, our cravings meant that those of us who got enough energy also obtained a balance of vitamins, minerals, and nonnutrients such as fiber and medicines. Processed foods meet our energy and protein needs without providing the same breadth of vitamins, minerals, fiber, and plant secondary compounds.[13] This is where we have gone wrong. The processed food industry supplies what we want, not what we need.

## TAKING BACK THE REINS

Fortunately, we humans retain much of our innate ability to select a healthy diet. Under experimental conditions, we prefer high-energy foods when hungry and low-energy foods when full. We eat more when a meal contains *varied* elements rather than just a single food, and we avoid new foods unless reassured by others. We avoid eating feces and the dead bodies of our own species. We retain remnants of those diet-selection strategies that help other species find a balanced diet while avoiding unfamiliar toxins and pathogens. When we are sick, we reject our food and often become irritable, desiring to be left alone. Nesse and Williams claim that we even avoid *specific* foods that would make our condition worse: "In the midst of a bout of influenza, such iron-rich foods as ham and eggs suddenly seem disgusting; we prefer tea and toast. This is just the ticket for keeping iron away from pathogens." We may even keep an ability to self-medicate by craving certain foods when suffering particular ills. Traditional Chinese medics have noticed that people start to crave soil and old tea leaves when infested with hookworms. Because the tea leaves are a concentrated source of tannins, this behavior is similar to that of other species (such as deer) that control internal parasites by eating tannin-rich plants. In America, African slaves infested with hookworms craved clay, and many individuals today have a desire to eat soil and clay when suffering gastrointestinal malaise. The medical profession has even developed a "self-medication hypothesis" to explain why some people drink alcohol, crave certain mood-modulating foods, and smoke nicotine or cannabis.[14]

Eating the right foods and natural medicines requires a sensitivity to subtle changes in appetite. Do I fancy something sweet, sour, salty, stimulating, or sedating? What sort of hunger is it? And, after consumption, has the "need" been satisfied? Such subtleties are easily overridden by artificially created superstimuli in processed foods that leave us unable to select a healthy diet. We need to listen more carefully to our body's cravings and take an intentional role in maintaining our health before disease sets in.

The ability of our species to design solutions to problems has been one of our strongest assets. Fantastic advances in surgical techniques and in pharmaceutical and genetic research have saved, and will continue to save, many lives. But it is important to set these developments (which are often profit-focused rather than person-focused) in a wider ecological context so that preventive health care is not devalued. The active self-help strategies that we have seen support wild health are not romantic idealizations of nature. They have emerged through the most ruthless test of efficacy: natural selection.

We need to know more about health ecology, yet if we are to learn more from observations of wild animals, we must protect them in their natural environment as complex interdependent ecosystems.

Only in this way can we see the full picture. If habitat destruction continues, and we allow pollution to undermine the health of our wildlife, we will lose a vast and valuable library of information. By protecting large areas of wilderness, we can conserve this essential resource for future generations.

# NOTES

## Introduction

1. D. Rockwell, *Giving Voice to Bear: North American Indian Myths, Rituals, and Images of the Bear* (Colorado: Roberts Rinehart, 1991); S. Sigstedt, unpublished AAAS proceedings, 1992; F. Densmore, *How Indians Use Wild Plants for Food, Medicine and Crafts* (New York: Dover Publications, 1928); D. C. Jarvis, *Folk Medicine: A Doctor's Guide to Good Health* (London: Carnell, 1957).
2. The most recent example is a BBC publication that accompanied the documentary series "Supernatural"; chimpanzee self-medication was discussed under the heading "Paranormal."

## 1. Health in the Wild

1. J. de Baïracli Levy, *The Complete Herbal Handbook for Farm and Stable* (London: Faber & Faber, 1984 [1952]), p. 10.
2. M. E. Roelke-Parker, L. Munson, et al., "A Canine Distemper Virus Epidemic in Serengeti Lions (*Panthera leo*)," *Nature*, 379(6564) (1996): 441–445; Africa News Service, New Vision (May 3, 2000); W. Conway, "Linking Zoo and Field, and Keeping Promises to Dodos," paper presented at the 7th World Conference on Breeding Endangered Species, Cincinnati Zoo, 1999.
3. "Sickly Swine," *New Scientist* (Sept. 25, 1999): 23.
4. "Public Enemy Number One," *New Scientist* (Oct. 2, 1999).
5. R. A. Mittermeier, N. Myers, et al., "Biodiversity and Major Tropical Wilderness Area: Approaches to Setting Conservation Priorities," *Conservation Biology*, 12(3) (1998): 516–520.
6. C. Moss, *Elephant Memories: Thirteen Years in the Life of an Elephant Family* (Chicago: University of Chicago Press, 1988 [2000]), p. 268.
7. G. B. Schaller, *The Deer and the Tiger: A Study of Wildlife in India* (Chicago: University of Chicago Press, 1967); R. M. Jakob-Hoff, "Diseases in Free-Living Marsupials," in *Zoo and Wild Animal Medicine*, ed. M. E. Fowler (Philadelphia: W. B. Saunders, 1993).

8. P. P. Calle, J. Rivas, et al., "Health Assessment of Free-ranging Anacondas (*Eunectes murinus*) in Venezuela," *Journal of Zoo and Wildlife Medicine*, 25(1) (1994): 53–62; W. B. Karesh, B. S. Alvaro de Campo, et al., "Health Evaluation of Free-ranging and Hand-reared Macaws (*Ara* sp.) in Peru," *Journal of Zoo and Wildlife Medicine*, 28(4) (1997): 368–377; K. V. K. Gilardi, L. J. Lowenstine, et al., "A Survey for Selected Viral Chlamydial and Parasitic Diseases in Wild Dusky-headed Parakeets (*Aratinga weddellii*) and Tui Parakeets (*Brotogeris sanctithomae* in Peru," *Journal of Wildlife Diseases*, 31(4) (1995): 523–528; W. B. Karesh, M. M. Uhart, et al., "Health Evaluation of Free-ranging Rockhopper Penguins (*Eudyptes chrysoco*)," *Journal of Zoo and Wildlife Medicine*, 30(1) (1999): 25–31; W. B. Karesh, M. M. Uhart, et al., "Health Evaluation of Free-ranging Guanaco (*Lama guanicoe*)," *Journal of Zoo and Wildlife Medicine*, 29(2) (1998): 34–141; William B. Karesh, *Appointment at the Ends of the World: Memoirs of a Wildlife Veterinarian* (New York: Warner Books, 2000); W. B. Karesh, A. Rothstein, et al., "Health Evaluation of Black-faced Impala (*Aepyceros melampus petersi*) Using Blood Chemistry and Serology," *Journal of Zoo and Wildlife Medicine*, 28(4) (1997): 361–367; B. L. Raphael, M.W. Klemens, et al., "Blood Values in Free-ranging Pancake Tortoises (*Malacochersus tornieri*)," *Journal of Zoo and Wildlife Medicine*, 25(1) (1994): 63–67.
9. Karesh, *Appointment at the Ends of the World*, p. 70.
10. B. L. Hart, "Behavioral Adaptations to Pathogens and Parasites: Five Strategies," *Neuroscience and Biobehavioral Reviews*, 14 (1990): 273–294.
11. M. A. Suleman, D. Yole, et al., "Peripheral Blood Lymphocyte Immunocompetence in Wild African Green Monkeys (*Cercopithecus aethiops*) and the Effects of Capture and Confinement," *In Vivo*, 1 (1999): 25–27.
12. J. E. Stevens, "The Delicate Art of Shark Keeping," *Sea Frontiers*, 41(1) (1995): 34–43.

## 2. Nature's Pharmacy

1. M. Wink, T. Schmeller, et al., "Modes of Action of Allelochemical Alkaloids: Interaction with Neuroreceptors, DNA and Other Molecular Targets," *Journal of Chemical Ecology*, 24(11) (1998): 1881–1937.
2. T. D. A. Forbes, I. J. Pemberton, et al., "Seasonal Variation of Two Phenolic Amines in *Acacia berlanieri*," *Journal of Arid Environments*, 30 (1995): 403–415; M. Jang and J. M. Pezzuto, "Cancer Chemopreventive Activity of Resveratrol," *Drug Experimentation and Clinical Research*, 25(2–3) (1999): 65–77.
3. J. B. Harborne, *Introduction to Ecological Biochemistry*, 4th ed. (London: Academic Press, 1993), pp. 245–247.
4. I. Shonle and J. Bergelson, "Interplant Communication Revisited," *Ecology*, 76(8) (1995): 2660–63; G. Cronin and M. E. Hay, "Induction of Seaweed Chemical Defences by Amphipod Grazing," *Ecology*, 77(8) (1996): 2287–2302.
5. J. Takabayahsi, M. Dicke, et al., "Leaf Age Affects Composition of Herbivore-

induced Synomones and Attraction of Predatory Mites," *Journal of Chemical Ecology,* 20 (1994): 373–386.

6. M. Robles, M. Aregullin, et al., "Recent Studies on the Zoopharmacognosy, Pharmacology and Neurotoxicology of Sesquiterpene Lactones," *Planta Medica,* 61 (1995): 199–203; W. C. Evans, *Trease and Evans' Pharmacognosy,* 13th ed. (Philadelphia: Baillière Tindall, 1989).

7. R. W. Wrangham and P. G. Waterman, "Condensed Tannins in Fruits Eaten by Chimpanzees," *Biotropica,* 15(3) (1993): 217–222.

8. United States Agricultural Research Services, phytochemical database managed by James Duke.

9. Evans, *Trease and Evans' Pharmacognosy;* M. I. Gabarev, E. Y. Enioutina, et al., "Plant-derived Glycoalkaloids Protect Mice Against Lethal Infection with *Salmonella typhimurium,*" *Phytotherapy Research,* 12(2) (1998): 79–88.

10. T. Elmqvist, R. G. Cates, et al., "Flowering in Males and Females of a Utah Willow, *Salix rigida,* and Effects on Growth, Tannins, Phenolics, Glycosides and Sugars," *Oikos,* 61 (1991): 65–72; R. Arvigo and M. Balik, *Rainforest Remedies: One Hundred Healing Herbs of Belize* (Lotus Press, 1993).

11. J. P. Bryant, P. J. Kuropat, et al., "Control over the Allocation of Resources by Woody Plant to Chemical Antiherbivore Defence," in *Plant Defences Against Mammalian Herbivores,* ed. R. T. Palo and C. T. Robbins (Ann Arbor, Mich.: CRC Press, 1991); R. T. Palo, J. Gowda, and P. Hogberg, "Species Height and Root Symbiosis: Two Factors Influencing Antiherbivore Defense of Woodyplants in East Africa Savanna," *Oecologia,* 93(3) (1993): 322–326; R. T. Palo, "Distribution of Birch (*Betula* sp.), Willow (*Salix* sp.) and Poplar (*Populus* sp.) Secondary Metabolites and Their Potential Role as Chemical Defense Against Herbivores," *Journal of Chemical Ecology,* 10 (1984): 499–520.

## 3. Food, Medicine, and Self-medication

1. H. C. Lu, *Chinese System of Food Cures: Prevention and Remedies* (New York: Sterling Publishing, 1986).

2. E. Giovannucci, "Tomatoes, Tomato-based Products, Lycopene, and Cancer: Review of the Epidemiologic Literature," *Journal of the National Cancer Institute,* 91(4) (1999): 317–331.

3. E. S. Dierenfeld, plenary lecture, symposium on Nutrition of Wild and Captive Wild Animals, *Proceedings of the Nutrition Society,* 56 (1997): 989–999.

4. S. J. Simpson and J. D. Abisgold, "Compensation by Locusts for Changes in Dietary Nutrients: Behavioural Mechanisms," *Physiological Entomology,* 10 (1985): 443–452.

5. E. J. Sterling, E. S. Dierenfeld, et al., "Dietary Intake, Food Composition, and Nutrient Intake in Wild and Captive Populations of *Daubentonia madagascariensis,*" *Folia Primatologica,* 62(1–3) (1994): 115–124; P. S. Simpkin of FarmAfrica, personal communication.

6. D. K. Thomas, "Figs as Food Source of Migrating Garden Warblers in Southern Portugal," *Bird Study,* 26 (1979): 187–191; F. Barlein and D. Simons, "Nu-

tritional Adaptations in Migrating Birds," *Israel Journal of Zoology,* 41 (1995): 357–367. There may be other reasons why figs are beneficial at this time — anthelmintic properties for example; C. L. Frank, E. S. Dierenfeld, and K. B. Storey, "The Relationship Between Lipid Pre-oxidation, Hibernation and Food Selection in Mammals," *American Zoologist,* 38(2) (1998): 341–349,

7  J. M. Bovee Oudenhoven, M. L. Wissink, et al., "Dietary Calcium Phosphate Stimulates Intestinal Lactobacilli and Decreases the Severity of a Salmonella Infection in Rats," American Society of Nutritional Science, *Biochemical and Molecular Action of Nutrients* (1999): 607–612.

8. S. J. McNaughton, "Mineral Nutrition and Spatial Concentrations of African Ungulates," *Nature,* 334(6180) (1988): 343–345; R. M. Jakob-Hoff, "Diseases in Free-living Marsupials," in *Zoo and Wild Animal Medicine,* ed. M. E. Fowler (London: W. B. Saunders, 1993).

9. R. W. Marlow and K. Tollestrope, "Mining and Exploitation of Natural Mineral Deposits by the Desert Tortoise, *Gopherus agassizii,*" *Animal Behaviour,* 30(2) (1982): 475–478.

10. B. Woodside and L. Millelire, "Self-selection of Calcium During Pregnancy and Lactation in Rats," *Physiology and Behavior,* 39 (1987): 291–295.

11. D. Western, "Giraffe Chewing a Grants Gazelle Carcass," *East African Wildlife Journal,* 9 (1971): 156–157.

12. R. Furness, "Predation on Ground-nesting Seabirds by Island Populations of Red Deer (*Cervus elaphus*) and Sheep (*Ovis*)," *Journal of Zoology, London,* 216 (1988): 565–573; F. D. Provenza, "Postingestive Feedback as an Elementary Determinant of Feed Preferences and Intake in Ruminants," *Journal of Range Management,* 48 (1995): 2–17.

13. J. R. Blairwest, D. A. Denton, et al., "Behavioural and Tissue Responses to Severe Phosphorus Depletion in Cattle," *American Journal of Physiology,* 263(3, pt. 2) (1992): 653–656.

14. W. B. Karesh, *Appointment at the Ends of the World* (New York: Warner Books, 1999), p. 188.

15. T. R. Hubback, "The Malay Elephant," *Journal of the Bombay Natural History Society,* 42 (1941): 483–509 (quotation from p. 506).

16. D. Bloch, "Salt and the Evolution of Money," *Journal of Salt History,* 7 (1999).

17. R. K. Siegel, *Intoxication: Life in Pursuit of Artificial Paradise* (New York: Pocket Books, 1989), p. 67.

18. Reno Sommerhalder, personal communication.

19. K. Nakamura and R. Norgren, "Sodium-deficient Diet Reduces Gustatory Activity in the Nucleus of the Solitary Tract of Behaving Rats," *American Journal of Physiology,* 269(3, pt. 2) (1995): 647–661; J. Schulkin, "The Allure of Salt," *Psychobiology,* 19(2) (1991): 116–121.

20. G. E. Belovsky, "Food Plant Selection by a Generalist Herbivore: The Moose," *Ecology,* 62 (1981): 1020–30.

21. J. A. FernandezLopez, M. Esteve, et al., "Management of Dietary Essential Metals (Iron, Copper, Zinc, Chromium and Manganese) by Wistar and Zucker Obese Rats Fed a Self-selected High-energy Diet," *Biometals,* 7(2) (1994): 117–129.

22. L. Watson, "The Biology of Being," in *Spirit of Science,* ed. D. Lorimer (Guildford, Sussex, U.K.: Floris Books, 1998), p. 167.

23. R. T. Cook, S. E. R. Bailey, et al., "The Influence of Nutritional Status on the Feeding Behaviour of the Field Slug (*Deroceras reticulatum*)," *Animal Behaviour,* 59 (2000): 167–176; J. J. Villalba and F. D. Provenza, "Postingestive Feedback from Starch Influences the Ingestive Behaviour of Sheep Consuming Wheat Straw," *Applied Animal Behaviour Science,* 66(1–2) (1999): 49–63; F. D. Provenza, "Post-ingestive Feedback as an Elementary Determinant of Food Preference and Intake in Ruminants," *Journal of Range Management,* 48 (1995): 2–17.

24. W. Rogers and P. Rozin, "Novel Food Preferences in Thiamine-deficient Rats," *Journal of Comparative Physiology and Psychology,* 61 (1996): 1–4; S. Feurte, S. Nicolaidis, and K. C. Berridge, "Conditioned Taste Aversion in Rats for a Threonine-deficient Diet: Demonstration by the Taste Reactivity Test," *Physiology and Behavior,* 68(3) (2000): 423–429.

25. M. Leshem, S. DelCanho, and J. Schulkin, "Calcium Hunger in the Parathyroidectomized Rat Is Specific," *Physiology and Behavior,* 67(4) (1999): 555–559; J. M. Sweeny, H. E. Seibert, et al., "Evidence for Induction of a Phosphate Appetite in Juvenile Rats," *American Journal of Physiology, Regulatory and Integrative and Comparative Physiology,* 44(4) (1998): R1358–65; Chris Mead, British Trust for Ornithologists, personal communication.

26. M. Boppré, "Redefining Pharmacophagy," *Journal of Chemical Ecology,* 10(7) (1984): 1151–54 (quotation from p. 1152).

27. E. D. Levin, W. Wilson, et al., "Nicotine-haloperidol Interactions and Cognitive Performance in Schizophrenics," *Neuropsychopharmacology,* 15(5) (1996): 429–436.

28. These criteria are discussed briefly in L. A. Ketch, D. Malloch, et al., "Comparative Microbial Analysis and Clay Mineralogy of Soils Eaten by Chimpanzees (*Pan troglodytes schweinfurthii*)," *Soil Biology and Biochemistry,* 33 (2001): 199–203.

## 4. Information for Survival

1. C. J. Anderson, "Animals, Earthquakes, and Eruptions," *Field Museum of Natural History Bulletin,* Chicago, 44(5) (1973): 9–11; Y. Suyehiro, "Unusual Behaviour of Fishes to Earthquakes," *Keikyu Aburatsubo Marine Park Aquarium Scientific Report,* 1 (1968): 4–11.

2. J. Adamson, *Queen of Shaba: The Story of an African Leopard* (London: Collins and Harvill Press, 1980); R. W. Marlow and K. Tollestrope, "Mining and Exploitation of Natural Mineral Deposits by the Desert Tortoise, *Gopherus agassizii*," *Animal Behaviour,* 30(2) (1982): 475–478.

3. J. Downer, *Supernatural: The Unseen Powers of Animals* (BBC Worldwide, 1999); W. R. Corliss, *Biological Anomalies: Birds* (U.S. Sourcebook Project, 1998), pp. 374–375.

4. B. Bagemihil, *Biological Exuberance: Animal Homosexuality and Natural Diversity* (New York: St. Martin's Press, 1999).

5. M. Roberts, *The Man Who Listens to Horses* (London: Hutchinson, 1996).
6. E. E. Shook, *Treatise on Advanced Herbology* (Pomeroy: Health Research, 1987), lesson 23, p. 1.
7. H. C. Lu, *Chinese Herbal Cures* (New York: Sterling Publishing, 1994 [1991]), p. 74; R. C. Wren, *Potter's New Cyclopedia of Botanical Drugs and Preparations* (Saffron Walden: C. W. Daniel, 1988 [1907]).
8. O. Lawson-Dick, ed., *Aubrey's Brief Lives* (Secker & Warburg, 1949).
9. D. Rockwell, *Giving Voice to Bear: North American Indian Myths, Rituals, and Images of the Bear* (Colorado: Roberts Rinehart, 1991).
10. M. A. Huffman, "Current Evidence for Self-medication in Primates: A Multidisciplinary Perspective," *Yearbook of Physical Anthropology,* 40 (1997): 171–200. Huffman is doing further research on this plant, and we shall soon hear of its potential in more detail.
11. Ricardo Leizaola, director of a documentary based on the life of Señor Benito Reyes (personal communication).
12. J. de Baïracli Levy, *Illustrated Herbal Handbook for Everyone* (London: Faber & Faber, 1974), p. 210; M. Mességué, *Of People and Plants: The Autobiography of Europe's Most Celebrated Herbal Healer* (Rochester, Vt.: Healing Arts Press, 1991), p. 13.
13. Although the experiments themselves may not be fatal, animals used in medical research usually must, by law, be sacrificed to prevent the escape of hazards into the community.
14. R. Lewontin, *It Ain't Necessarily So: The Dream of the Human Genome and Other Illusions* (London: Granta Books, 2000), p. 267.
15. J. Goodall, *Through a Window: My Thirty Years with the Chimpanzees of Gombe* (London: Weidenfeld & Nicolson, 1990), p. 19.
16. M. A. Huffman and M. K. Seifu, "Observations on the Illness and Consumption of a Possibly Medicinal Plant *Vernonia amygdalina* by Wild Chimpanzees in the Mahale Mountains National Park, Tanzania," *Primates,* 30(1) (1989): 51–63.
17. Reprinted with additional comments in R. M. Sapolsky, *Junk Food Monkeys* (London: Headline, 1998), pp. 151–166.

## 5. Poisons

1. M. L. Drew and M. E. Fowler, "Poisoning of Black and White Ruffed Lemurs (*Varecia variegata variegata*) by Hairy Nightshade (*Solanum sarrachoides*)," *Journal of Zoological Wildlife Medicine,* 22(4) (1991): 494–496.
2. M. E. Fowler, "Plant Poisoning in Free-living Wild Animals: A Review," *Journal of Wildlife Diseases,* 19(1) (1983): 34–43.
3. D. S. Shepard and I. R. Inglis, "Toxic Bait Aversions in Different Rat Strains Exposed to an Acute Rodenticide," *Journal of Wildlife Management,* 57(3) (1993): 640–647; J. W. Noble, J. D. Crossley, et al., "Pyrrolizidine Alkaloidosis of Cattle Associated with *Senecio lautus*," *Australian Veterinary Journal,* 71(7) (1994): 196–200.
4. M. Viette, C. Tettamanti, and F. Saucy, "Preference for Acyanogenic White

Clover (*Trifolium repens*) in the Vole (*Arvicola terrestris*): II. Generalizations and Further Investigations," *Journal of Chemical Ecology*, 26(1) (2000): 101–122.

5. This idea has been proposed by many authors, notably Timothy Johns. It should be noted that some of these detection mechanisms may be transferable to other nonfamiliar toxins.

6. A. G. Chiarello, "Diet of the Atlantic Forest Maned Sloth (*Bradypus torquatus*)," *Journal of Zoology*, 246(1) (1998): 11–19.

7. B. D. Loutit, G. N. Louw, and M. K. Seely, "First Approximation of Food Preferences and the Chemical Composition of the Diet of the Desert-dwelling Black Rhinoceros, *Diceros bicornis*," *Madoqua*, 15 (1987): 35–54.

8. B. G. Galef, Jr., and E. E. Whiskin, "Social Influences on the Amount of Food Eaten by Norway Rats," *Appetite*, 34(3) (2000): 327–332; B. G. Galef, Jr., "Direct and Indirect Behavioral Pathways to the Social Transmission of Food Avoidance," *Annals of the New York Academy of Sciences*, 443 (1984): 203–215.

9. P. G. Hepper, "Adaptive Fetal Learning: Prenatal Exposure to Garlic Affects Postnatal Preference," *Animal Behaviour*, 36 (1988): 935–936; W. P. Smotherman, "Odor Aversion Learning by the Rat Fetus," *Physiology and Behavior*, 29 (1982): 769–771; S. N. Mirza and F. D. Provenza, "Socially Induced Food Avoidance in Lambs — Direct or Indirect Maternal Influence," *Journal of Animal Science*, 72(4) (1994): 899–902.

10. T. Johns, *The Origins of Human Diet and Medicine* (Tucson: University of Arizona Press, 1993), p. 40.

11. R. A. Distel and F. D. Provenza, "Experience Early in Life Affects Voluntary Intake of Blackbrush by Goats," *Journal of Chemical Ecology*, 17(2) (1991): 431–450; J. B. Harborne, *Introduction to Ecological Biochemistry*, 4th ed. (London: Academic Press, 1993); J. A. Pfister, F. D. Provenza, et al., "Tall Larkspur Ingestion: Can Cattle Regulate Intake Below Toxic Levels?" *Journal of Chemical Ecology*, 23(3) (1997): 759–777.

12. E. J. Calabrese and L. A. Baldwin, "Hormesis as a Biological Hypothesis," *Environmental Health Perspectives*, 106(S1) (1998): 357–362.

13. E. M. Fowler, "Plant Poisoning in Free-living Wild Animals: A Review," *Journal of Wildlife Diseases*, 19(1) (1983): 34–43 (quotation from p. 40).

14. T. L. Hailey, J. W. Thomas, and R. M. Robinson, "Pronghorn Die-off in Trans-Pecos, Texas (Tar Brush *Fluorensia cernua*)," *Journal of Wildlife Management*, 30 (1966): 488–496; N. B. Metcalfe, F. A. Huntingford, and J. E. Thorpe, "Predation Risk Impairs Diet Selection in Juvenile Salmon," *Animal Behaviour*, 35(3) (1987): 931–933.

15. W. J. Freeland, P. H. Calcott, and L. R. Anderson, "Tannins and Saponins: Interaction in Herbivore Diets," *Biochemical Systematics and Ecology*, 13(2) (1981): 189–193.

16. F. D. Provenza, E. A. Buritt, et al., "Self-regulation of Polyethylene Glycol by Sheep Fed Diets Varying in Tannin Concentrations," *Journal of Animal Science*, 78(5) (2000): 1206–12.

17. J. F. Oates, "The Guereza and Its Food," in *Primate Ecology: Studies of Feeding and Ranging Behaviour in Lemurs, Monkeys and Apes*, ed. T. H. Clutton-

Brock (London: Academic Press, 1977), pp. 275–321; W. S. Goldstein and K. C. Spencer, "Inhibition of Cyanogenesis by Tannins," *Journal of Chemical Ecology*, 11(7) (1985): 847–858.

18. Johns, *Origins of Human Diet and Medicine*.

19. E. Visalberghi and E. Addessi, "Response to Changes in Food Palatability in Tufted Capuchin Monkeys (*Cebus apella*)," *Animal Behaviour*, 59 (2000): 231–238.

20. M. D. Dearing, "The Manipulation of Plant Toxins by a Food-hoarding Herbivore (*Ochotona princeps*)," *Ecology*, 78(3) (1997): 774–781.

21. P. N. Newton, "The Ecology and Social Organisation of Hanuman Langurs (*Presbytis entellus*) in Kanha Tiger Reserve, Central Indian Highlands," (D.Phil. diss., Oxford University, 1984).

22. P. W. Abrahams and J. A. Parsons, "Geophagy in the Tropics: A Literature Review," *Geographical Journal*, 162(1) (1996): 63–73.

23. K. Izawa, "Soil-eating by *Alouatta* and *Ateles*," *International Journal of Primatology*, 14(2) (1993): 229–242; C. M. Hladik, "A Comparative Study of the Feeding Strategies of Two Sympatric Species of Leaf Monkeys: *Presbytis senex* and *Presbytis entellus*," in Clutton-Brock, *Primate Ecology*, pp. 324–353; W. J. Foley and C. Macarthur, "The Effects and Costs of Allelochemicals for Mammalian Herbivores: An Ecological Perspective," in *The Digestive Systems of Mammals: Food, Form and Function*, ed. D. J. Chivers and P. Langer (Cambridge: Cambridge University Press, 1994).

24. W. E. Faber, A. Pehrson, and P. A. Jordan, "Seasonal Use of Salt Blocks by Mountain Hares in Sweden," *Journal of Wildlife Management*, 57(4) (1993): 842–846; C. Sun, T. C. Moermond, and T. J. Givnish, "Nutritional Determinants of Diet in Three Turacos in a Tropical Montane Forest," *Auk*, 114(2) (1997): 200–211.

25. I. Redmond, "Salt Mining Elephants of Mount Elgon," *Swara*, 5 (1982): 28–31.

26. J. F. Oates, "Water-plant and Soil Consumption by Guereza Monkeys (*Colobus guereza*): A Relationship with Minerals and Toxins in the Diet?" *BioTropica*, 10(4) (1978): 241–253.

27. Abrahams and Parsons, "Geophagy in the Tropics."

28. D. A. Kruelen, "Lick Use by Large Herbivores: A Review of Benefits and Banes of Soil Consumption," *Mammal Reviews*, 15(3) (1985): 107–123.

29. G. B. Schaller, *The Year of the Gorilla* (Chicago: University of Chicago Press, 1964); D. Fossey, *Gorillas in the Mist* (London: Hodder & Stoughton, 1983).

30. W. C. Mahaney, S. Aufreiter, and R. G. V. Hancock, "Mountain Gorilla Geophagy: A Possible Seasonal Behavior for Dealing with the Effects of Dietary Changes," *International Journal of Primatology*, 16(3) (1995): 475–488.

31. W. C. Mahaney, R. G. V. Hancock, et al., "Geochemistry and Clay Mineralogy of Termite Mound Soils and the Role of Geophagy in Chimpanzees of Mahale Mountains, Tanzania," *Primates*, 37(2) (1996): 121–134. For a review of geophagy in primates, see R. Krishnamani and W. C. Mahaney, "Geophagy Among Primates: Adaptive Significance and Ecological Consequences," *Animal Behaviour*, 59 (2000): 899–915.

32. G. Klaus, C. Klaus-Hugi, and B. Schmid, "Geophagy by Large Mammals at Natural Licks in the Rain Forest of the Dzanga National Park, Central African Republic," *Journal of Tropical Ecology,* 14 (1998): 829–839.

33. Discussed in W. Mayer, "Feat of Clay," *Wildlife Conservation* (June 1999).

34. J. D. Gilardi, S. S. Duffey, et al., "Biochemical Functions of Geophagy in Parrots: Detoxification of Dietary Toxins and Cytoprotective Effects," *Journal of Chemical Ecology,* 25(4) (1999): 897–919.

35. N. Takeda, S. Hasegawa, et al., "Pica in Rats Is Analogous to Emesis: An Animal Model in Emesis Research," *Pharmacology Biochemistry and Behavior,* 45(4) (1993): 817–821; R. M. Sapolsky, *Junk Food Monkeys* (London: Headline, 1998) p. 156.

36. In his book *The Deer and the Tiger: A Study of Wildlife in India* (Chicago: University of Chicago Press, 1967), Schaller refers to this observation in a paper by Powell (1957); wolf scats containing earth, David Mech (personal communication).

37. K. A. Bolton, V. M. Campbell, et al., "Chemical Analysis of Soils of Kowloon (Hong Kong) Eaten by Hybrid Macaques," *Journal of Chemical Ecology,* 24(2) (1998): 195–205.

38. J. de Baïracli Levy, *The Complete Herbal Handbook for Farm and Stable* (London: Faber & Faber, 1984 [1952]), p. 151.

39. S. T. Tyler, "The Behaviour and Social Organisation of the New Forest Pony," *Animal Behaviour Monographs,* 5 (1972): 96.

40. T. T. Struhsaker, D. O. Cooney, and K. S. Siex, "Charcoal Consumption by Zanzibar Red Colobus Monkeys: Its Function and Its Ecological and Demographic Consequences," *International Journal of Primatology,* 18(1) (1997): 61–72; D. O. Cooney and T. T. Struhsaker, "Adsorptive Capacity of Charcoals Eaten by Zanzibar Red Colobus Monkeys: Implications for Reducing Dietary Toxins," *International Journal of Primatology,* 18(2) (1997): 235–246.

41. J. Davenport, *Environmental Stress and Behavioural Adaptation* (London: Croom Helm, 1985).

42. R. D. Fairshter and A. F. Wilson, "Paraquat Poisoning: Manifestations and Therapy," *American Journal of Medicine,* 59(6) (1975): 751–753.

## 6. Microscopic Foes

1. R. Root-Bernstein and M. Root-Bernstein, *Honey, Mud, Maggots, and Other Medical Marvels* (Boston: Houghton Mifflin, Mariner Books, 1998).

2. Strictly speaking, antibiotics are produced only by living organisms such as fungi, but the term is commonly used to denote any substances that kill bacteria. Pathogens can be organisms of any size, but in this chapter I use the word to refer to the commonly held perception of "germs" that are invisible to the naked eye.

3. R. M. Sapolsky, *Why Zebras Don't Get Ulcers: An Updated Guide to Stress, Stress-related Diseases, and Coping* (New York: W. H. Freeman, 1998[1994]).

4. R. M. Jakob-Hoff, "Diseases in Free-Living Marsupials," in *Zoo and Wild Animal Medicine,* ed. M. E. Fowler (London: W. B. Saunders, 1993).

5. The varroa mite is an ectoparasite rather than a bacterium or virus. O. Boeking, "Sealing Up and Non-removal of Diseased and *Varroa jacobsoni* Infested Drone Brood Cells Is Part of the Hygienic Behaviour in *Apis cerana*," *Journal of Apicultural Research*, 38(3–4) (1999): 159–168.

6. J. Goodall, *The Chimpanzees of Gombe: Patterns of Behavior* (Cambridge, Mass: Belknap Press of Harvard University Press, 1986), p. 543. For a general review of how animals avoid disease see C. Loehle, "Social Barriers to Pathogen Transmission in Wild Animal Populations," *Ecology*, 76(2) (1995): 326–376.

7. Survival International (personal communication).

8. D. W. Pfennig, S. G. Ho, and E. A. Hoffman, "Pathogen Transmission as a Selective Force Against Cannibalism," *Animal Behaviour*, 55(5) (1998): 1255–61.

9. M. Parker-Pearson, "The Archaeology of Death and Burial" (Texas A & M University Anthropology Series, no. 3, 2001 [1999]).

10. M. R. Hacker, B. A. Rothernburg, and M. J. Kluger, "Plasma Iron, Copper, and Zinc in Lizard *Diposaurus dorsalis:* Effects of Bacterial Injection," *American Journal of Physiology*, 240(5) (1999): R272–275; W. W. Reynolds, "Behavioural Fever in Teleost Fishes," *Nature*, 259 (1976): 41–42.

11. S. Blanford, M. B. Thomas, and J. Langewald, "Behavioural Fever in the Senegalese Grasshopper (*Oedaleus senegalensis*) and Its Implications for Biological Control Using Pathogens," *Ecological Entomology*, 23(1) (1998): 9–14; P. T. Starks, C. A. Blackie, and T. D. Seely, "Fever in Honeybee Colonies," *Naturwissenschaften*, 87 (2000): 229–231.

12. R. M. Nesse and G. C. Williams, *Why We Get Sick: The New Science of Darwinian Medicine* (New York: Vintage Books, 1994), pp. 27–29.

13. J. de Baïracli Levy, *The Complete Herbal Handbook for Farm and Stable* (London: Faber & Faber, 1984 [1952]), p. 351.

14. Nesse and Williams, *Why We Get Sick*, pp. 36–39.

15. M. J. Han, H. Y. Park, et al., "Protective Effects of *Bifodobacterium* sp. on Experimental Colon Carcinogenesis with 1,2-dimethylhydrazine," *Journal of Microbiological Biotechnology*, 9(3) (1999): 368–370; I. M. Bovee-Oudenhoven, M. L. Wissink, et al., "Dietary Calcium Phosphate Stimulates Intestinal Lactobacilli and Decreases the Severity of a Salmonella Infection in Rats," American Society of Nutritional Science, *Biochemical and Molecular Action of Nutrients* (1999): 607–612.

16. D. S. Newburg and J. M. Street, "Bioactive Materials in Human Milk: Milk Sugars Sweeten the Argument for Breast-feeding," *Nutrition Today*, 32(5) (1997): 191–202.

17. S. T. Tyler, "The Behaviour and Social Organisation of the New Forest Pony," *Animal Behaviour Monographs*, 5 (1972).

18. H. S. Costa, T. J. Henneberry, and N. C. Toscano, "Effects of Antibacterial Material on *Bemisia argentifolii* (Homoptera: Aleyrodidae) Oviposition, Growth, Survival, and Sex Ratio," *Journal of Economic Entomology*, 90(2) (1997): 333–339.

19. J. P. Berry, "Chemical Ecology of Mountain Gorillas (*Gorilla gorilla beringei*) with Special Reference to Antimicrobial Constituents in the Diet," Ph.D diss., Cornell University, Ithaca, N.Y., 1998.

20. J. D. Hare and T. G. Andreadis, "Variation in the Susceptibility of *Leptinotarsa decemlineata* (Coleoptera: Chrysomelidae) When Reared on Different Host Plants to the Fungal Pathogen *Beauveria bassiana* in the Field and Laboratory," *Environmental Entomology*, 12 (1983): 1891–96; R. K. Chandra, "Influence of Palm Oil on Immune Responses and Susceptibility to Infection in a Mouse Model," *Nutrition Research*, 16(1) (1996): 61–68.

21. M. L. Gubarev, M. L. Enioutina, et al., "Plant-derived Glycoalkaloids Protect Mice Against Lethal Infection with *Salmonella typhimurium*," *Phytotherapy Research*, 12 (1998): 79–88.

22. R. Voelker, "The Hygiene Hypothesis," *Journal of the American Medical Association*, 283(10) (2000): 1282; S. Kleiner, "The Hygiene Hypothesis Gains Further Momentum in Childhood Asthma," *Lancet*, 355(9217) (2000): 1795.

23. M. Chipako, "Propolis: The Beeswax That Heals," *Africa News* (Mar. 12, 1998).

24. M. L. Taper and T. I. Case, "Interactions Between Oak Tannins and Parasite Community Structure: Unexpected Benefits of Tannins to Cynipid Gall Wasps," *Oecologia*, 71(2) (1987): 254–261.

25. D. W. Tallamy, "Sequestered Cucurbitacins and Pathogenicity of *Metarhizium anisopliae* (Moniliales: Moniliaceae) on Spotted Cucumber Beetle Eggs and Larvae (Coleoptera: Chrysomelidae)," *Environmental Entomology*, 27(2) (1998): 366–372; Y. Wang, U. G. Mueller, and J. Clardy, "Antifungal Dikeropiperazines from Symbiotic Fungus of Fungus Growing Ant (*Cyphomyrax minutes*)," *Journal of Chemical Ecology*, 25(4) (1999): 935–941. The fungus prevents other fungi from infecting the ant's food.

26. P. Proksch, "Defensive Roles for Secondary Metabolites from Marine Sponges and Sponge-feeding Nudibranchs," *Toxicon*, 32(6) (1994): 639–655.

27. V. A. Krischik, P. Barbosa, and C. F. Reicheklderfer, "Three Trophic Levels Interactions: Allelochemicals, *Manduca Sexta* (L) and *Bacillus thuringiensis* var. *Kurstaki* Berliner," *Environmental Entomology*, 17 (1988): 476–482.

28. J. de Baïracli Levy, *The Complete Herbal Handbook for Farm and Stable* (London: Faber & Faber, 1984 [1952]).

29. W. C. Mahaney, J. Zippin, et al., "Chemistry, Mineralogy and Microbiology of Termite Oils Eaten by the Chimpanzees of the Mahale Mountains, Western Tanzania," *Journal of Tropical Ecology*, 15 (1999): 565–588; L. A. Ketch, D. Malloch, et al., "Comparative Microbial Analysis and Clay Mineralogy of Soils Eaten by Chimpanzees (*Pan troglodytes schweinfurthii*)," *Soil Biology and Biochemistry*, 33 (2001): 199–203.

30. F. Vincent, "Utilisation spontanée d'outils chez le mandrill (primate)," *Mammalia*, 37 (1973): 277–280.

31. T. Nishida and M. Nankamura, "Chimpanzee Tool Use to Clear a Blocked Nasal Passage," *Folia Primatologia*, 61 (1993): 218–220.

32. N. Kakiuchi, M. Hattori, et al., "Studies on Dental Caries Prevention by Tra-

ditional Medicines. 8. Inhibitory Effect of Various Tannins on Glucan Synthesis by Glucosyltransferase from *Streptococcus mutans,*" *Chemical Pharmacology Bulletin,* 34(2) (1986): 720–725.

33. W. C. McCrew and C. E. G. Tutin, "Chimpanzee Tool Use in Dental Grooming," *Nature,* 241 (1973): 477–478.

34. H. Neu, "The Crisis in Antibiotic Resistance," *Science* 257 (1992): 1064–73; M. Hakkinen and C. Schneitz, "Efficacy of a Commercial Competitive Exclusion Product Against a Chicken Pathogenic *Escherichia coli* and *E. coli* 0157:H7," *Veterinary Record,* 139 (1997): 139–141; A. Coglan, "Antibiotic from Gut Bug Keeps Killer at Bay," *New Scientist* (Feb. 15, 1997).

35. I. T. Kudva, P. G. Hatfield, and C. J. Hovde, "Effect of Diet on the Shedding of *Escherichia coli* 0157-H7 in a Sheep Model," *Applied and Environmental Microbiology,* 61(4) (1995): 1363–70; F. Diez-Gonzalez, T. R. Callaway, et al., "Grain Feeding and the Dissemination of Acid-Resistant *E. coli* from Cattle," *Science,* 281(5383) (Sept. 11, 1998): 1666–68; D. D. Hancock, T. E. Besser, et al., "Cattle, Hay, and *E. coli,*" *Science* 284(5411) (Apr. 2, 1999): 51–52.


## 7. Gaping Wounds and Broken Bones

1. D. Fossey, *Gorillas in the Mist* (London: Hodder & Stoughton, 1983), p. 69.
2. C. Moss, *Portraits in the Wild: Animal Behaviour in East Africa* (London: Hamish Hamilton, 1976), p. 109.
3. A. Jolly, *The Evolution of Primate Behavior,* 2nd ed. (New York: Macmillan, 1985); G. Wobeser, "Traumatic Degenerative and Developmental Lesions in Wolves and Coyotes from Saskatchewan," *Journal of Wildlife Diseases,* 28(2) (1992): 268–275.
4. J. Goodall, *The Chimpanzees of Gombe: Patterns of Behavior* (Cambridge, Mass.: Belknap Press of Harvard University Press, 1986), p. 100.
5. T. de Almeida, *Jaguar Hunting: In the Matto Grosso and Bolivia* (Safari Press, 1990), p. 50.
6. C. Drews, "Contexts and Patterns of Injuries in Free-ranging Male Baboons (*Papio cynocephalus*)," *Behaviour,* 133 (1996): 443–474.
7. C. Drews, "Road Kills of Animals by Public Traffic in Mikumi National Park, Tanzania, with Notes on Baboon Mortality," *African Journal of Ecology,* 33(2) (1995): 89–100.
8. B. E. Rollin, *The Unheeded Cry: Animal Consciousness, Animal Pain and Scientific Change,* expanded ed. (Iowa State University Press, Priority Press, 1988).
9. F. C. Colpaert, P. Dewitte, et al., "Self-administration of the Analgesic Suprofen in Arthritic Rats: Evidence of *Mycobacterium butyricum*-induced Arthritis as an Experimental Model of Chronic Pain," *Life Sciences,* 27 (1980): 921–928; R. Kupers and J. Gybels, "The Consumption of Fentanyl Is Increased in Rats with Nociceptive Pain but Not with Neuropathic Pain," *Pain,* 60 (1995): 137–141.
10. T. C. Danbury, C. A. Weeks, et al., "Self-selection of the Analgesic Drug Carprofen by Lame Broiler Chickens," *Veterinary Record* (Mar. 11, 2000).

11. D. Mech, *The Wolf: The Ecology and Behavior of an Endangered Species* (New York: Natural History Press, 1970).
12. C. Moss, *Elephant Memories: Thirteen Years in the Life of an Elephant Family* (Chicago: University of Chicago Press, 2000 [1988]), p. 261.
13. Fossey, *Gorillas in the Mist*, p. 84.
14. B. L. Hart, "Behavioral Adaptations to Pathogens and Parasites: Five Strategies," *Neuroscience and Biobehavioral Reviews*, 14 (1990): 273–294.
15. J. M. Hutson, M. Niall, et al., "Effect of Salivary Glands on Wound Contraction in Mice," *Nature*, 279 (1979): 793–795.
16. See review in R. Root-Bernstein and M. Root-Bernstein, *Honey, Mud, Maggots, and Other Medical Marvels* (Boston: Houghton Mifflin, Mariner Books, 1998), pp. 110–118.
17. S. A. Foster, "Wound Healing: A Possible Role of Cleaning Stations," *Copeia*, 4, (1985): 875–880.
18. D. P. Reid, *Chinese Herbal Medicine* (Boston: Shambala, 1993).
19. P. J. Hudson, *The Moor: Nature's Healing Miracle of Health, Rejuvenation and Beauty* (Eastbourne: Mayfair Publishing, 1993).
20. B. Brinkhaus, M. Linder, et al., "Chemical, Pharmacological and Clinical Profile of the East Asian Medical Plant *Centella asiatica*," *Phytomedicine*, 7(5) (2000): 427–448.
21. G. Schenk, *The Book of Poisons*, trans. M. Bullock (New York: Rinehart, 1955), p. 52.
22. Fossey, *Gorillas in the Mist*, p. 173.
23. B. G. Ritchie and M. Fragaszy, "Capuchin Monkey (*Cebus apella*) Grooms Her Infant's Wound with Tools," *American Journal of Primatology*, 16 (1988): 345–348; G. Westergaard and D. Fragaszy, "Self-treatment of Wounds by a Capuchin Monkey (*Cebus apella*)," *Human Evolution*, 1(6) (1987): 557–562.
24. Ernest Thompson Seton wrote widely on these matters in his nonfiction accounts of natural history; bear observations from Gilles Roanich (personal communication); R. Dexreit, *Our Earth, Our Cure: A Handbook of Natural Medicine for Today* (Citadel Press, 1993); deer example from R. Siegel, *Intoxication: Life in Pursuit of Artificial Paradise* (New York: Pocket Books, 1989), p. 42.
25. I. Douglas-Hamilton and O. Douglas-Hamilton, *Among the Elephants* (London: Collins & Harvill, 1975).
26. T. R. Hubback, "The Malayan Elephant," *Journal of the Bombay Natural History Society*, 42 (1941): 483–509; K. Payne, *Silent Thunder: The Hidden Voice of Elephants* (London: Weidenfeld & Nicolson, 1998), p. 23.
27. M. Mességué, *Of People and Plants: The Autobiography of Europe's Most Celebrated Herbal Healer* (Rochester, Vt.: Healing Arts Press, 1991), p. 14; L. Watson, "The Biology of Being," in *The Spirit of Science: From Experiments to Experience*, ed. D. Lorimer (Guildford, Sussex, U.K.: Floris Books, 1998), p. 167.
28. W. J. Long, "Animal Surgery," *The Outlook* (Sept. 12, 1903); "New Pastor in 'Nature War'," *Tribune* (June 12, 1907).
29. J. Goodall, *Through a Window: My Thirty Years with the Chimpanzees of Gombe* (London: Weidenfeld & Nicolson, 1990), p. 98.

30. Fossey, *Gorillas in the Mist*, p. 93.
31. O. A. E. Rasa, "A Case of Invalid Care in Wild Dwarf Mongooses," *Zeitschrift für Tierpsychologie*, 62 (1983): 235–240.
32. Payne, *Silent Thunder*, p. 53.
33. E. E. Shook, *Advanced Treatise on Herbology* (Pomeroy: Health Research, 1987), p. 8; R. R. Chilpa and M. J. Estrada, "Chemistry of Antidotal Plants," *Interciencia*, 20(5) (1995): 257.
34. E. Wigginton, *The Foxfire Book* (New York: Anchor Press, 1972), pp. 229–230.
35. R. R. Swaisgood, D. H. Owings, and M. P. Rowe, "Conflict and Assessment in a Predator-prey System: Ground Squirrels Versus Rattlesnakes," *Animal Behaviour*, 57(5) (1999): 1033–44.
36. F. Markland and Q. Zhow, "Snake Venom Disintegrin: An Effective Inhibitor of Breast Cancer Growth and Dissemination," *Natural and Selected Synthetic Toxins*, 745 (2000): 262–282.

### 8. Mites, Bites, and Itches

1. A. Rabinowitz, *Jaguar: One Man's Struggle to Establish the World's First Jaguar Preserve* (Washington, D.C.: Island Press, 2000 [1986]), p. 24.
2. B. L. Hart, "Behavioral Adaptations to Pathogens and Parasites: Five Strategies," *Neuroscience and Biobehavioral Reviews*, 14 (1990): 223–294.
3. A. P. Möller, "The Preening Activity of Swallows *Hirundo rustica* in Relation to Experimentally Manipulated Loads of Haematophagous Mites," *Animal Behaviour*, 42 (1991): 251–260.
4. Survival International (personal communication).
5. T. R. Hubback, "The Malayan Elephant," *Journal of the Bombay Natural History Society*, 42 (1941): 483–509; B. L. Hart and L. A. Hart, "Fly Switching by Asian Elephants: Tool Use to Control Parasites," *Animal Behaviour*, 48 (1994): 35–45.
6. Reviewed in B. L. Hart, "Behavioural Defense Against Parasites: Interaction with Parasite Invasiveness," *Parasitology*, 109 (1994): S139–S151.
7. J. M. Butler and T. J. Roper, "Ectoparasites and Sett Use in European Badgers," *Animal Behaviour*, 52 (1996): 621–629.
8. L. A. Hart and B. L. Hart, "Autogrooming and Social Grooming in Impala," *Annals of the New York Academy of Sciences*, 525 (1988): 399–402; P. Christie, H. Richner, and A. Oppliger, "Of Great Tits and Fleas: Sleep Baby Sleep," *Animal Behaviour*, 52 (1996): 1087–92.
9. J. Goodall, *Through a Window: My Thirty Years with the Chimpanzees of Gombe* (London: Weidenfeld & Nicolson, 1990), p. 142.
10. P. Stopka and D. W. MacDonald, "The Market Effect in the Wood Mouse (*Apodemus sylvaticus*): Selling Information on Reproductive Status," *Ethology*, 105(11) (1999): 960–982; M. de L. Brooke, "The Effect of Allo-preening on Tick Burdens of Molting Eudyptid Penguins," *Auk*, 102 (1985): 893–895.
11. H. M. Feder, "Cleaning Symbiosis in the Marine Environment," in S. M. Henry, ed., *Symbiosis*, vol. 1 (New York: Academic Press, 1966), pp. 327–380;

A. Bjordal, "Wrasse as Cleaner Fish for Farmed Salmon," *Progress in Underwater Science*, 16 (1991): 17–28.

12. H. Greene, *Snakes: The Evolution of Mystery in Nature* (Berkeley: University of California Press, 1997).

13. M. Baker, "Fur-rubbing: Use of Medicinal Plants by Capuchin Monkeys (*Cebus capuchinus*)," *American Journal of Primatology*, 38(3) (1996): 263–270; quotation from interview recorded in "Go Ape over Primates," *Ranger Rick* (National Wildlife Federation, Oct. 1997).

14. E. Neal and C. Cheeseman, *Badgers* (London: Poyser Natural History, 1996).

15. X. Valderrama, J. G. Robinson, et al., "Seasonal Anointment with Millipedes in a Wild Primate: A Chemical Defense Against Insects?" *Journal of Chemical Ecology*, 26(12) (2000): 2781–90.

16. K. R. Gibson, "Tool Use, Imitation and Deception in Captive Cebus Monkey," in *Language and Intelligence in Monkeys and Apes*, ed. S. T. Parker and K. R. Gibson (Cambridge: Cambridge University Press, 1990), pp. 205–218.

17. M. E. Gompper and A. M. Holyman, "Grooming with *Trattinnickia* Resin: Possible Pharmaceutical Plant Use by Coatis in Panama," *Journal of Tropical Ecology*, 9 (1997): 533–540.

18. M. Aregullin, A. Pelayo, and E. Rodriguez, "Terpenes from the Resin of *Trattinnickia aspera* (Burseraceae) Used in Grooming by Coatis" (in prep.); M. Robles, M. Arguellin, et al., "Recent Studies on the Zoopharmacognosy, Pharmacology and Neurotoxicology of Sesquiterpene Lactones," *Planta Medica*, 61(6) (1995): 199.

19. S. Sigstedt, unpublished AAAS proceedings, 1992. Discussed in E. Rodriguez and R. W. Wrangham, "Zoopharmacognosy: The Use of Medicinal Plants by Animals," in *Phytochemical Potential of Tropical Plants*, ed. K. R. Downum et al. (New York: Plenum Press, 1993).

20. S. G. Gillespie and J. N. Duszynski, "Phthalides and Monoterpenes of the Hexane Extracts of the Roots of *Ligusticum porteri, L. filicinum und L. tenulifolium*," *Planta Medica*, 64(4) (1998): 392; D. Moerman, *Native American Ethnobotany* (Portland, Ore.: Timber Press, 1998); Gilles Roanen, personal communication.

21. J. Coats and C. Peterson, "Catnip and Osage Orange Components Found to Repel German Cockroaches," paper presented at American Chemical Society national meeting, New Orleans, Aug. 1999.

22. F. Brinker, "The Insecticidal and Therapeutic Activity of Natural Isobutylamides," *British Journal of Phytotherapy*, 2(4) (1991/2): 160–170.

23. D. C. Hauser, "Anting by Gray Squirrels," *Journal of Mammology*, 45 (1964): 136–138; J. T. Longino, "True Anting by the Capuchin *Cebus capuchinus*," *Primates*, 25(2) (1984): 243–245.

24. D. H. Clayton and N. D. Wolfe, "The Adaptive Significance of Self-medication," *Trends in Ecology and Evolution*, 8(2) (1993): 60–61.

25. J. Kochansky and H. Shimanuki, "Development of a Gel Formulation of Formic Acid for Control of Parasitic Mites of Honey Bees," *Journal of Agricultural and Food Chemistry*, 47(9) (1999): 3850–58.

26. B. Furlow, "Kills All Known Germs," *New Scientist* (Jan. 22, 2000).
27. D. Clayton and J. Vernon, "Common Grackle Anting with Lime Fruit and Its Effect on Ectoparasites," *Auk,* 110(4) (1993): 951–952.
28. For a review of the many ways birds manage ectoparasites, see B. R. Moyer and D. H. Clayton, "Avian Defenses Against Ectoparasites," in *Insect and Bird Interactions,* ed. H. F. Emden and M. Rothschild (Andover, U.K. 2001).
29. Information supplied by S. Piers Simpkin of FarmAfrica.
30. E. W. Heymann, "Urine-washing and Related Behaviour in Wild Moustached Tamarins, *Saguinus mystax* (Callitrichidae)," *Primates,* 36(2) (1995): 259–264.
31. R. Root-Bernstein and M. Root-Bernstein, *Honey, Mud, Maggots, and Other Medical Marvels* (Boston: Houghton Mifflin, Mariner Books, 1998).
32. M. Grieves, *A Modern Herbal* (London: Tiger Books, 1931), p. 162.
33. L. Clark and J. R. Mason, "Effect of Biologically Active Plants Used as Nest Material and the Derived Benefit to Starling Nestlings," *Oecologia,* 77 (1988): 174–180.
34. L. Clark and J. R. Mason, "Use of Nest Material as Insecticidal and Anti-pathogenic Agents by the European Starling," *Oecologia* (Berlin), 67 (1985): 169–176.
35. L. Clark and J. R. Mason, "Olfactory Discrimination of Plant Volatiles by the European Starling," *Animal Behaviour,* 35 (1987): 227–235; reviewed in T. J. Roper, "Olfaction in Birds," *Advances in the Study of Behavior,* 28 (1999): 247–332 (p. 305).
36. P. T. Fauth, D. G. Krementz, and J. E. Hines, "Ectoparasitism and the Role of Green Nesting Material in the European Starling," *Oecologia,* 88(1) (1991): 22–29; M. Eens, R. Pinxten, and R. F. Verheyen, "Function of the Song and Song Repertoire in the European Starling (*Sturnus vulgaris*): An Aviary Experiment," *Behaviour,* 125(1–2) (1993): 51–66.
37. H. Gwinner, M. Oltrongge, et al., "Green Plants in Starling Nests: Effects on Nestlings," *Animal Behaviour,* 59(2) (2000): 301–309.
38. Hawk research described in M. J. Plotkin, *Medicine Quest: In Search of Nature's Healing Secrets* (New York: Viking Press, 2000), p. 174; S. Senegupta and Shrilata, "House Sparrow *Passer domesticus* Uses Krishnachura Leaves as an Antidote to Malarial Fever," *Emu,* 97 (1997): 248.
39. Reviewed in Moyer and Clayton, "Avian Defenses Against Ectoparasites."
40. Neal and Cheeseman, *Badgers;* Grieves, *A Modern Herbal.*
41. T. Eitz, W. M. Whitten, et al., "Fragrance Collection, Storage and Accumulation by Individual Male Orchid Bees," *Journal of Chemical Ecology,* 25(1) (1999): 157–176.
42. J. P. Dumbacher, "Evolution of Toxicity in Pitohuis: I. Effects of Homobatrachotoxin on Chewing Lice (order Phthiraptera)," *Auk,* 116 (1999): 957–962; J. Louis-Jaques, M. Aregullin, et al., "Chemical Analysis of Uropygial Gland Secretions of Three Neo-tropical Bird Species," abstract presented at undergraduate research poster session, Cornell University, Ithaca, N.Y., 1999.
43. A range of new breakthroughs is discussed in M. Knott, "Mean and Minty,"

*New Scientist* (Nov. 20, 1999): 22; W. S. Bowers, "Phytochemical Defences Targeting Insect Behaviour," *Acta Botanica Gallica,* 144(4) (1997): 383–390.

## 9. Reluctant Hosts, Unwelcome Guests

1. J. Goodall, *The Chimpanzees of Gombe: Patterns of Behavior* (Cambridge, Mass.: Belknap Press of Harvard University Press, 1986), p. 96; G. B. Schaller, *The Year of the Gorilla* (Chicago: University of Chicago Press, 1997 [1965]), p. 190; B. L. Raphael, M. W. Klemens, et al., "Blood Values in Free-ranging Pancake Tortoises (*Malacochersus tornieri*)," *Journal of Zoo and Wildlife Medicine,* 25(1) (1994): 63–67; E. Neal and C. Cheeseman, *Badgers* (London: Poyser Natural History, 1996); W. B. Karesh, M. M. Uhart, et al., "Health Evaluation of Free-ranging Rockhopper Penguins (*Eudyptes chrysoco*)," *Journal of Zoo and Wildlife Medicine,* 30(1) (1999): 25–31.
2. B. L. Hart, "Behavioural Defence Against Parasites: Interaction with Parasite Invasiveness," *Parasitology,* 109 (1994): S139–S151.
3. Goodall, *Chimpanzees of Gombe,* p. 545.
4. See review by B. L. Hart, "Behavioral Adaptations to Parasites: Five Strategies," *Neuroscience and Biobehavioral Reviews,* 14 (1990): 273–294.
5. J. O. Evans, S. Piers Simpkin, and D. J. Atkins, "Camel Keeping in Kenya," *Range Management Handbook,* 3(8) Ministry of Agriculture, Livestock, Development and Marketing, Kenya, 1995. This relatively new idea is proposed by Hart, "Behavioural Defence," and G. A. Lonzano, "Optimal Foraging Theory: A Possible Role for Parasites," *Oikos,* 60 (1991): 391–395.
6. D. H. Janzen, "Complications in Interpreting the Chemical Defences of Trees Against Tropical Arboreal Plant-eating Vertebrates," in *The Ecology of Arboreal Folivores,* ed. G. G. Montgomeries (Washington, D.C.: Smithsonian Institution Press, 1978), pp. 73–84.
7. J. D. Kabasa, J. OpudaAslbo, and U. TerMeulen, "The Effect of Oral Administration of Polyethylene Glycol on Faecal Helminth Egg Counts in Pregnant Goats Grazed on Browse Containing Condensed Tannins," *Tropical Animal Health Production,* 32(2) (2000): 73–86; H. VerheydenTixier and P. Duncan, "Selection for Small Amounts of Hydrolysable Tannins by a Concentrate Selecting Mammalian Herbivore," *Journal of Chemical Ecology,* 26(2) (2000): 351–358; S. O. Hoskin, T. N. Barry, et al., "Effects of Reducing Anthelmintic Input upon Growth and Faecal Egg and Larval Counts in Young Farmed Deer Grazing Chicory (*Cichorium intybus*) and Perennial Ryegrass (*Lolium perenne* White Clover (*Trifolium repens* Pasture," *Journal of Agricultural Science,* 132(3) (1999): 335–345.
8. J. E. Phillips-Conroy, "Baboons, Diet and Disease: Food Plant Selection and Schistosomiasis," in *Current Perspectives in Primate Social Dynamics,* ed. D. M. Taub and F. A. King (New York: Von Nostrand Reinhold, 1986), pp. 287–304.
9. M. Galal, A. K. Bashir, et al., "Efficacy of Aqueous and Butanol Fractions of *Albizzia anthelmintica* Against Experimental *Hymnolepis diminuta* Infestation in Rats," *Veterinary and Human Toxicology,* 33(6) (1991): 537–539.

10. S. Piers Simpkin of FarmAfrica, personal communication.
11. R. Burton, "The Tiger as Fruit Eater," *Journal of the Bombay Natural History Society*, 50 (1952): 649; G. B. Schaller, *The Deer and the Tiger: A Study of Wildlife in India* (Chicago: University of Chicago Press, 1967), p. 280.
12. K. E. Glander, "Nonhuman Primate Self-medication with Wild Plant Foods," in *Eating on the Wild Side: The Pharmacologic, Ecologic, and Social Implications of Using Noncultigens*, ed. Nina L. Etkin (Tucson: University of Arizona Press, 1994), pp. 227–239; M. J. Chatton, *Handbook of Medical Treatments* (Los Altos, Calif.: Lange Medical Publications, 1972).
13. M. D. Stuart, K. B. Strier, and S. M. Pierberg, "A Coprological Survey of Wild Muriquis *Brachyteles arachnoids* and Brown Howling Monkeys *Aloutta fusca*," *Journal of the Helminthological Society* (1993).
14. O. Courtenay, "Conservation of the Maned Wolf: Fruitful Relationships in a Changing Environment," *Canid News*, 2; L. Munson and R. J. Montali, "High Prevalence of Ovarian Tumors in Maned Wolves (*Chrysocyon brachyurus*) at the National Zoological Park," *Journal of Zoo and Wildlife Medicine*, 22(1) (1991): 125–129; E. K. P. da Silveira, "O. lobo-guara (*Chyrsocyon brachyrus*). Possival Acao Inhibidoria de Certas Solancas Sobre o Nematoide Renal," *Vellozia*, 1 (1969): 58–60.
15. A. A. Forsyth, *British Poisonous Plants* (London: Ministry of Agriculture, Fisheries and Food, bulletin no. 61, 1954).
16. R. W. Wrangham field notes, personal communication.
17. E. Rodriguez, M. Aregullin, et al., "Thiarubrin A, a Bioactive Constituent of *Aspilia* Consumed by Wild Chimpanzees," *Experientia*, 41 (1985): 419–420.
18. See review by R. W. Wrangham and J. Goodall, "Chimpanzee Use of Medicinal Leaves," in *Understanding Chimpanzees*, ed. P. G. Heltne and L. A. Marquardt (Cambridge, Mass.: Harvard University Press, 1989), pp. 22–37.
19. P. N. Newton and T. Nishida, "Possible Buccal Administration of Herbal Drugs by Wild Chimpanzees, *Pan troglodytes*," *Animal Behaviour*, 39(4) (1990): 799–800.
20. E. Rodriguez and R. W. Wrangham, "Zoopharmacognosy: The Use of Medicinal Plants by Animals," *Recent Advances in Phytochemistry*, 27: "Phytochemical Potential of Tropical Plants," ed. K. R. Downum, J. T. Romeo, and H. Stafford (New York: Plenum Press, 1993), pp. 89–105.
21. M. Kawabata and T. Nishida, "A Preliminary Note on the Intestinal Parasites of Wild Chimpanzees in the Mahale Mountains, Tanzania," *Primates*, 32 (2) (1991): 275–278; J. E. Page, M. A. Huffman, et al., "Chemical Basis for Medicinal Consumption of *Aspilia* (Asteraceae) Leaves by Chimpanzees: A Reanalysis," *Journal of Chemical Ecology*, 23(9) (1997): 2211–25.
22. E. J. Messer and R. W. Wrangham, "In-vitro Testing of the Biological Activity of *Rubia cordifolia* Leaves on Primate *Strongyloide* Species," *Primates*, 37(1) (1995): 105–108.
23. See review in M. A. Huffman, "Current Evidence for Self-medication in Primates: A Multidisciplinary Perspective," *Yearbook of Physical Anthropology*, 40 (1997): 171–200.

24. R. W. Wrangham, "Leaf Swallowing by Chimpanzees, and Its Relation to a Tapeworm Infection," *American Journal of Primatology*, 37 (1995): 297–303.
25. M. A. Huffman and J. M. Caton, "Self-induced Gut Motility and the Control of Parasite Infections in Wild Chimpanzees," *International Journal of Primatology*, 22(3) (2001): 329–346.
26. Ibid.
27. Ibid.; M. Mességué, *Of People and Plants: The Autobiography of Europe's Most Celebrated Herbal Healer* (Rochester, Vt.: Healing Arts Press, 1991), pp. 13–14.
28. Tiger examples are cited in Schaller, *The Deer and the Tiger*, p. 280; wolf observations in A. Murie, *The Wolves of Mount McKinley* (Washington, D.C.: U.S. Department of the Interior, Fauna Series 5, 1944), p. 59.
29. M. A. Huffman and M. K. Seifu, "Observations on the Illness and Consumption of a Possibly Medicinal Plant *Vernonia amygdalina* by a Wild Chimpanzee in the Mahale Mountains National Park, Tanzania," *Primates*, 30(1) (1989): 51–63; A. Hofer, M. A. Huffman, and G. Ziesler, *Mahale: A Photographic Encounter with Chimpanzees* (New York: Sterling Publishing, 2000).
30. K. Koshimizu, H. Ohigashi, and M. A. Huffman, "Use of *Vernonia amygdalina* by Wild Chimpanzees: Possible Roles of Its Bitter and Related Compounds," *Physiology and Behavior*, 56(6) (1994): 1209–16.
31. M. A. Huffman, S. Gotoh, et al., "Seasonal Trends in Intestinal Nematode Infection and Medicinal Plant Use Among Chimpanzees in the Mahale Mountains, Tanzania," *Primates*, 38(2) (1997): 111–125.
32. M. A. Huffman, H. Ohigashi, et al., "African Great Ape Self-medication: A New Paradigm for Treating Parasite Disease with Natural Medicines?" in *Towards Natural Medicine in the Twenty-first Century*, ed. H. Ageta, N. Ami, et al. (Elsevier Science, 1998), pp. 113–123.
33. Koshimizu, Ohigashi, and Huffman, "Use of *Vernonia amygdalina*"; R. M. Sapolsky, *Junk Food Monkeys* (London: Headline, 1998), p. 160.
34. S. Vitazkova, E. Long, et al., "Mice Suppress Malaria Infection by Sampling a 'Bitter' Chemotherapy Agent," *Animal Behaviour*, 61(5) (2001): 887–894.
35. M. McRae, "Creature Cures," *Equinox*, 75 (May/June 1994).
36. M. A. Huffman, "Practical Applications from the Study of Great Ape Self-medication and Conservation Related Issues," *Pan Africa News*, 4(2) (1997): 15–16.
37. W. C. Campbell and S. S. Duffey, "Tomatine and Parasitic Wasps: Potential Incompatibility of Plant Antibiosis with Biological Control," *Science*, 205 (1979): 700–702.
38. R. Karban and G. English-Loeb, "Tachinid Parasitoids Affect Host Plant Choice by Caterpillars to Increase Caterpillar Survival," *Ecology*, 78(2) (1997): 603–611.
39. See review in J. Moore, "The Behavior of Parasitized Animals," *Bioscience*, 45(2) (1995): 89–98.
40. J. de Baïracli Levy, *The Complete Herbal Handbook for Farm and Stable* (London: Faber & Faber, 1984 [1952]), p. 167.

41. J. M. Hunter, "Geophagy in Africa and the US: A Culture-infestation Hypothesis," *Geographical Reviews*, 63 (1973): 173–195.
42. M. Knezevich, "Geophagy as a Therapeutic Mediator of Endoparasitism in a Free-ranging Group of Rhesus Macaques (*Macaca mulatta*)," *American Journal of Primatology*, 44(1) (1998): 71–82.
43. D. E. Elliot, J. F. Urban, et al., "Does the Failure to Acquire Helminthic Parasites Predispose to Crohn's Disease!" *Journal of the Federation of American Societies of Experimental Biology*, 14(12) (2000): 1848–55.

## 10. Getting High

1. E. Marais, cited in R. K. Siegel (see note 2); N. Gordon, "Monkey Business," *BBC Wildlife* (June 2001): 22–28; S. D. Fitzgerald, J. M. Sullivan, and R. J. Everson, "Suspected Ethanol Toxicosis in Two Wild Cedar Waxwings," *Avian Diseases*, 34 (1990): 488–490.
2. Ronald K. Siegel uncovered many examples of animal intoxication during his exploration of human drug abuse. A number of examples discussed in this chapter come from his book, *Intoxication: Life in Pursuit of Artificial Paradise* (published by Pocket Books in 1989 and now, sadly, out of print).
3. T. Morner and C. H. Segerstad, "Winter Mortality in Waxwings (*Bombycilla garrulous*) Caused by Ethanol Intoxication," *Journal of Wildlife Diseases* (in press).
4. "Elephants Rampage, Trample 5 in India," *Los Angeles Times* (Jan. 1, 1985), p. 19.
5. W. H. Drummond, *The Large Game and Natural History of South and Southeast Africa* (Edinburgh: Hamilton, 1875).
6. Siegel, *Intoxication*, p. 118.
7. R. M. Palmour, J. Mulligan, et al., "Of Monkeys and Men: Vervets and the Genetics of Human-like Behaviors," *American Journal of Human Genetics*, 61 (1997): 481.
8. "Stress Hormone Linked to Increased Alcohol Consumption in Animal Model," *National Institute on Alcohol Abuse and Alcoholism* (May 22, 2000).
9. R. K. Siegel and M. Brodie, "Alcohol Self-administration by Elephants," *Bulletin of the Psychonomic Society*, 22(1) (1984): 49–52.
10. R. Dudley, "Evolutionary Origins of Human Alcoholism in Primate Frugivory," *Quarterly Review of Biology*, 75 (2000): 3–15.
11. B. A. McMillen and H. L. William, "Role of Taste and Calories in the Selection of Ethanol by C57BL/6NHsd and Hsd:ICR Mice," *Alcohol*, 15(3) (1998): 193–198.
12. E. B. Rimm, P. Williams, et al., "Moderate Alcohol Intake and Lower Risk of Coronary Heart Disease: Meta-analysis of Effects on Lipids and Haemostatic Factors," *British Medical Journal*, 319 (1999): 1523–28; I. Sample, "Your Good Health," *New Scientist* (Nov. 13, 1999); R. Dudley, "Fermenting Fruit and the Historical Ecology of Ethanol Ingestion: Is Alcoholism in Modern Humans an Evolutionary Hangover?" (in prep.).

13. H. G. Pope, "*Tabernathe iboga:* An African Narcotic Plant of Social Importance," *Economic Botany,* 23 (1969): 174–184.
14. J. Ott, "The Delphic Bee: Bees and Toxic Honeys as Pointers to Psychoactive and Other Medicinal Plants," *Economic Botany,* 52(3) (1998): 260–266.
15. T. R. Hubback, "The Malayan Elephant," *Journal of the Bombay Natural History Society,* 42 (1941): 483–509 (quotations from p. 490).
16. M. J. Plotkin, *Medicine Quest: In Search of Nature's Healing Secrets* (New York: Viking Press, 2000).
17. W. H. Hamilton, R. E. Bushkirk, and W. H. Bushkirk, "Omnivory and Utilization of Food Resources by Chacma Baboons (*Papio ursinus*)," *American Naturalist,* 112(987) (1978): 911–924; Siegel, *Intoxication,* pp. 24–25.
18. E. Rodriguez and J. C. Cavin, "The Possible Role of Amazonian Psychoactive Plants in the Chemotherapy of Parasitic Worms — A Hypothesis," *Journal of Ethnopharmacy,* 6 (1982): 303–309; J. Weiskopf, "From Agony to Ecstasy: The Transformative Spirit of Yaje," *Shaman's Drum* (autumn 1995).
19. H. van Lawick and J. van Lawick-Goodall, *Innocent Killers* (London: Collins, 1976 [1970]), p. 123.
20. Reviewed in A. Kuzmin, S. Semenova, et al., "Enhancement of Morphine Self-administration in Drug-naïve, Inbred Strains of Mice by Acute Emotional Stress," *European Neuropsychopharmacology,* 6 (1996): 63–68.
21. A. Zimmer, A. M. Zimmer, et al., "Increased Mortality, Hyperactivity and Hypoalgesia in Cannabinoid CB1 Receptor Knockout Mice," *Neurobiology,* 96(10) (1999): 5780–85; V. Peralta and M. J. Cuesta, "Influence of Cannabis Abuse on Schizophrenic Psychopathology," *Acta Psychiatrica Scandinavica,* 85(2) (1992): 127–130.
22. P. Leyhausen, "Addictive Behaviour in Free-ranging Animals," in *Psychic Dependence,* ed. L. Goldberg and F. Hoffmeister (Heidelberg: Springer-Verlag, 1972).
23. A. A. Forsyth, *British Poisonous Plants* (London: Ministry of Agriculture, Fisheries and Food, HMSO, bulletin no. 61, 1954).
24. P. Popik and P. Skolnick, "Pharmacology of Ibogaine and Ibogaine-related Alkaloids," in *The Alkaloids,* ed. G. A. Cordell (New York: Academic Press, 1999), vol. 52, pp. 197–231.

## 11. Psychological Ills

1. T. Hall, *To the Elephant Graveyard* (London: John Murray, 2000).
2. D. F. McAlpine, "Common Eider (*Somateria mollisima*) Incubates Gadwell (*Anas strepera* Eggs: A Case of Clutch Adoption due to Human Disturbance," *Canadian Field Naturalist,* 110(4) (1996): 707–708.
3. J. Goodall, *Through a Window: My Thirty Years with the Chimpanzees of Gombe* (London: Weidenfeld & Nicolson, 1990).
4. B. Wechsler, "Coping and Coping Strategies: A Behavioural Review," *Applied Animal Behaviour Science,* 43 (1995): 123–134.
5. For a thorough and accessible description of stress, immunity, and psycho-

logical health in humans and other mammals, see R. M. Sapolsky, *Why Zebras Don't Get Ulcers: An Updated Guide to Stress, Stress-related Diseases, and Coping* (New York: W. H. Freeman, 1998).

6. D. L. Castles and A. Whiten, "Post-conflict Behaviour of Wild Olive Baboons: I. Reconciliation, Redirection and Consolation," *Ethology,* 104(2) (1998): 126–147; and II. "Stress and Self-directed Behaviour," pp. 148–160.

7. S. L. Watson, J. P. Ward, et al., "Scent-marking and Cortisol Response in the Small-eared Bushbaby (*Otolemur garnettii*)," *Physiology and Behavior,* 66(4) (1999): 695–699; A. Schilling, "Olfactory Communication in Prosimians," in *The Study of Prosimian Behavior,* ed. G. A. Doyle and A. C. Walker (New York: Academic Press, 1979), pp. 347–363.

8. A. Kuzmin, S. Semenova, et al., "Enhancement of Morphine Self-administration in Drug-naïve, Inbred Strains of Mice by Acute Emotional Stress," *European Neuropsychopharmacology,* 6 (1996): 63–68.

9. N. F. Ramsey and J. M. Van Ree, "Emotional but Not Physical Stress Enhances Intravenous Cocaine Self-administration in Drug-naïve Rats," *Brain Research,* 608(2) (1993): 216–222.

10. A. V. Shlyahova and T. M. Vorobyova, "Control of Emotional Behaviour Based on Biological Feedback," *Neurophysiology,* 31(1) (1999): 38–40.

11. See review in F. Bellisle, J. E. Blundell, et al., "Functional Food Science and Behaviour and Psychological Functions," *British Journal of Nutrition,* 80(1) (1998): 173–193.

12. P. Levine, *Waking the Tiger: Healing Trauma, the Innate Capacity to Transform Overwhelming Experiences* (North Atlantic Books, 1997).

13. B. Beerda, M. B. H. Schilder, et al., "Behavioural and Hormonal Indicators of Enduring Environmental Stress in Dogs," *Animal Welfare,* 9(1) (2000): 49–62; R. VandenBos, "Post-conflict Stress Response in Confined Group Living Cats (*Felis silvestris catus*)," *Applied Animal Behaviour Science,* 59(4) (1998): 323–330.

14. A. K. Dixon, "Ethological Strategies for Defence in Animals and Humans: Their Role in Some Psychiatric Disorders," *British Journal of Medical Psychology* 71(4) (1998): 417–445; H. van Lawick and J. van Lawick-Goodall, *Innocent Killers* (London: Collins, 1976 [1970]), p. 19.

15. Goodall, *Through a Window,* pp. 175–176.

16. B. Thierry, L. Steru, et al., "Searching Waiting Strategy — A Candidate for an Evolutionary Model of Depression," *Behavioural and Neural Biology,* 41(2) (1984): 180–189; R. M. Nesse and G. C. Williams, *Why We Get Sick* (New York: Vintage Books, 1996), p. 209.

17. K. E. Lukas, "A Review of Nutritional and Motivational Factors Contributing to the Performance of Regurgitation and Reingestion in Captive Lowland Gorillas (*Gorilla gorilla gorilla*)," *Applied Animal Behaviour Science,* 63 (1999): 237–249; B. Wechsler and I. Schmid, "Aggressive Pecking by Males in Breeding Groups of Japanese Quail (*Coturnix japonica*)," *British Poultry Science,* 39(3) (1998): 333–339; S. S. da Cunha Nogueira, S. L. G. Nogueira-Filho, et al., "Determination of the Causes of Infanticide in Capybara (*Hy-*

*drochaeris hydrochaeris*) Groups in Captivity," *Applied Animal Behaviour Science,* 62 (4) (1999): 351–357.

18. S. W. Hansen, B. Houbak, and J. Malmkvist, "Development and Possible Causes of Fur Damage in Farm Mink — Significance of Social Environment," *Acta Agriculturae Scandinavica,* section A, Animal Science, 48(1) (1998): 58–64.

19. BBC news report of World Health Organization conference October 27, 1999; Institute of Psychiatry in London, and Harvard Medical School; T. L. Leaman, "Anxiety Disorders," *Primary Care,* 26(2) (1999): 197; World Health Organization on-line database; D. Morris, *The Human Zoo* (London: Jonathan Cape, 1994 [1969]), p. vii.

## 12. Family Planning

1. D. A. Shutt, "The Effect of Plant Oestrogens on Animal Reproduction." *Endeavour,* 75 (1976): 110–113.

2 . S. E. Rickard and L. U. Thompson, "Phytoestrogens and Lignans: Effects on Reproduction and Chronic Disease," *ACS Symposium Series,* 662 (1997): 273–293.

3. C. Sonnenschein and A. M. Soto, "An Updated Review of Environmental Estrogen and Androgen Mimics and Antagonists," *Journal of Steroid Biochemistry and Molecular Biology,* 65(1–6) (1998): 143–150.

4. F. T. Halaweish, D. W. Tallamy, and E. Santana, "Cucurbitacins: A Role in Cucumber Beetle Steroid Nutrition?" *Journal of Chemical Ecology,* 25(10) (1999): 2373–83.

5. J. B. Harborne, *Introduction to Ecological Biochemistry,* 4th ed. (London: Academic Press, 1993).

6. P. J. Berger, N. C. Negus, et al., "Chemical Triggering of Reproduction in *Microtus montanus,*" *Science,* 214(4516) (1981): 69–70.

7. J. D. Garey, "A Possible Role for Secondary Plant Compounds in the Regulation of Primate Breeding Cycles," *American Journal of Physical Anthropology,* 63(2) (1984): 160 (abstracts).

8. P. L. Whitten, "Flowers, Fertility, and Females," *American Journal of Physical Anthropology,* 60(2) (1983): 269–270 (abstracts); J. D. Garey, L. Markiewicz, and E. Gurpide, "Estrogenic Flowers, a Stimulus Foraging Activity in Female Vervet Monkeys," in *Proceedings of the 14th Congress of the International Primatological Society* (Strasbourg: IPS, 1992), p. 210.

9. K. B. Strier, "Menu for a Monkey," *Natural History,* 3 (1993): 34–42.

10. J. W. Bok, L. Lermer, et al., "Antitumor Sterols from the Mycelia of *Cordyceps sinensis,*" *Phytochemistry,* 51(7) (1999): 891–898; J. S. Zhu, G. M. Halpern, and K. Jones, "The Scientific Rediscovery of a Precious Ancient Chinese Herbal Regimen: *Cordyceps sinensis,* part II," *Journal of Alternative and Complementary Medicine,* 4(4) (1998): 429–457; M. J. Plotkin, *Medicine Quest: In Search of Nature's Healing Secrets* (New York: Viking Press, 2000), p. 46.

11. H. C. Lu, *Chinese Herbal Cures* (New York: Sterling Publishing, 1994 [1991]), p. 121.
12. A. Marchlewskakoj, "Sociogenic Stress and Rodent Reproduction," *Neuroscience and Biobehavioral Reviews*, 21(5) (1997): 699–703; L. M. Westlin and S. M. Ferreria, "Do Pouched Mice Alter Litter Size Through Resorption in Response to Resource Availability?" *South African Journal of Wildlife Research*, 30(3) (2000): 118–121.
13. A. L. Hughes, "Female Infanticide: Sex Ratio Manipulation in Humans," *Ethology and Socio-biology*, 2(3) (1981): 109–111.
14. I. Nishiumi, S. Yamagishi, et al., "Paternal Expenditure Is Related to Brood Sex Ratio in Polygynous Great Reed Warblers," *Behavioural Ecology and Sociobiology*, 39(4) (1996): 211–217.
15. T. H. Clutton-Brock, S. D. Albon, and F. E. Guinness, "Maternal Dominance, Breeding Success, and Birth Sex Ratios in Red Deer," *Nature*, 308 (1984): 358–360.
16. R. J. Deslippe and R. Savolainen, "Sex Investment in a Social Insect — The Proximate Role of Food," *Ecology*, 76 (2) (1995): 375–382; D. B. Meikle and M. W. Thornton, "Premating and Gestational Effects of Maternal Nutrition on Secondary Sex Ratio in House Mice," *Journal of Reproduction and Fertility*, 105(2) (1995): 193–196.
17. D. Prevedelli and R. Z. Vandini, "Survival and Sex Ratio of *Dinophilus gyrociliatus* (Polychaeta: doniphilidae) Under Different Dietary Conditions," *Marine Biology*, 133(2) (1999): 231–236; E. Z. Cameron, W. L. Linklater, et al., "Birth Sex Ratios Relate to Mare Condition at Conception in Kaimanawa Horses," *Behavioural Ecology*, 10(5) (1999): 472–475.
18. K. E. Glander, "Nonhuman Primate Self-medication with Wild Plant Foods," in *Eating on the Wild Side: The Pharmacologic, Ecologic, and Social Implications of Using Noncultigens*, ed. N. L. Etkin (Tucson: University of Arizona Press, 1994), pp. 227–239.
19. J. D. Garey, L. Erb Meuli, et al., "Diet and Twinning in Rhesus Monkeys," *American Journal of Primatology*, 8(4) (1985): 338–339 (abstracts).
20. J. A. R. A. M. VanHooff, "The Socio-ecology of Sex Ratio Variation in Primates: Evolutionary Deduction and Empirical Evidence," *Applied Animal Behaviour Science*, 51(3–4) (1997): 293–306.
21. R. Kilner, "Primary and Secondary Sex Ratio Manipulation by Zebra Finches," *Animal Behaivour*, 56 (1998): 155–164.
22. J. V. Gedir and R. J. Hudson, "Seasonal Foraging Behaviour Compensation in Reproductive Wapiti Hinds (*Cervus elaphus canadensis*)," *Applied Animal Behaviour Science*, 67(1–2) (2000): 137–150; D. Carlson and J. Griffin, *The Dog Owner's Home Veterinary Handbook* (Howell Book House, 1992), p. 328; D. Bewley, "Healing Meals," *BBC Wildlife* (Sept. 1997).
23. A. Jolly, *The Evolution of Primate Behavior*, 2nd ed. (New York: Macmillan, 1985), p. 79.
24. H. Dublin, personal communication; M. A. Huffman, personal communication.
25. M. L. Sauther, "Wild Plant Use by Pregnant and Lactating Ringtailed Lemurs,

with Implications for Early Hominid Foraging," in *Eating on the Wild Side: The Pharmacologic, Ecologic, and Social Implications of Using Noncultigens,* ed. N. L. Etkin (Tucson: University of Arizona Press, 1994), pp. 240–256.

26. B. D. Worden, P. C. Parker, and P. W. Pappas, "Parasites Reduce Attractiveness and Reproductive Success in Male Grain Beetles," *Animal Behaviour,* 59(3) (2000): 543–550.

27. M. Boppré, "Pharmacophagy in Adult Lepidoptera: The Diversity of a Syndrome," in *Tropical Biodiversity and Systematics,* Proceedings of 1997 international symposium on Biodiversity and Systematics in Tropical Ecosystems, Bonn, pp. 285–289.

28. B. L. Hart, E. K. Korinek, and P. L. Brennan, "Post-copulatory Grooming in Male Rats Prevents Sexually-transmitted Diseases," *Annals of the New York Academy of Sciences,* 525 (1988): 397–398.

## 13. Facing the Inevitable

1. J. Goodall, *The Chimpanzees of Gombe: Patterns of Behavior* (Cambridge, Mass.: Belknap Press of Harvard University Press, 1986).

2. H. van Lawick and J. van Lawick-Goodall, *Innocent Killers* (London: Collins, 1976 [1970]), p. 153.

3. G. Wobeser, "Traumatic Degenerative and Developmental Lesions in Wolves and Coyotes from Saskatchewan," *Journal of Wildlife Diseases,* 28(2) (1992): 268–275; C. E. L. Cantley, E. C. Firth, et al., "Naturally Occurring Osteoarthritis in the Metacarpophalangeal Joints of Wild Horses," *Equine Veterinary Journal,* 31(1) (1999): 73–81. An overview is provided in R. Nowak, "A Healthy Mind: Why Are Great Apes Resistant to the Ravages of Dementia?" *New Scientist,* (Jan. 27, 2001); S. Deguise, D. Martineau, et al., "Possible Mechanism of Action of Environmental Contaminants on St. Lawrence Beluga Whales (*Delphinapterus leucas*)," *Environmental Health Perspectives,* 103 (1995): 73–77.

4. R. M. Nesse and G. C. Williams, *Why We Get Sick* (New York: Vintage Books, 1996), p. 114.

5. D. P. Shanely and T. B. L. Kirkwood, "Calorie Restriction and Aging: A Life-history Analysis," *Evolution,* 54(3) (2000): 740–750.

6. J. Hutcher, "Mortality in the 15th Century," *Economic History Reviews,* 39(1) (1986): 9–38; World Health Organization 1999 data.

7. P. G. Xiao, S. T. Xing, and L. W. Wang, "Immunological Aspects of Chinese Medicinal Plants and Antiaging Drugs," *Journal of Ethnopharmacology,* 38(2–3) (1993): 167–175.

8. J. A. Joseph, B. Shukitt-Hale, et al., "Reversals of Age-related Declines in Neuronal Signal Transduction Cognitive and Motor Behaviour Deficits with Blueberry, Spinach or Strawberry Dietary Supplementation," *Journal of Neuroscience,* 19(18) (1999): 8114–21.

9. "Looking Out for Badly Behaved Elephants," *Royal Geographic Society Magazine,* 72(6) (2000): 83.

10. C. Moss, *Portraits in the Wild: Animal Behaviour in East Africa* (London:

Hamish Hamilton, 1976), p. 253; M. A. Huffman, personal communication; M. A. Huffman, "Some Socio-behavioural Manifestations of Old Age," in *The Chimpanzees of the Mahale Mountains: Sexual and Life History Strategies*, ed. T. Nishida (Tokyo: University of Tokyo Press, 1990), pp. 235–255.

11. R. M. Saplosky, *Why Zebras Don't Get Ulcers* (New York: W. H. Freeman, 1998), p. 320.
12. D. Fossey, *Gorillas in the Mist* (London: Hodder & Stoughton, 1983), p. 155.
13. Originally reported by ornithologist R. C. Murphy of the American Museum of Natural History in 1913, and later reviewed in W. R. Corliss, *Biological Anomalies: Birds* (U.S. Sourcebook Project, 1998), Section BBB41.
14. B. Thorson, "Boom and Bust (Lemmings' Life Cycle)," *Canadian Geographic*, 118(2) (1998): 68 –74.
15. R. K. Siegel, *Intoxication: Life in Pursuit of Artificial Paradise* (New York: Pocket Books, 1989), p. 42.
16. Described in H. Williams, *Sacred Elephant* (London: Jonathan Cape, 1989), p. 143.
17. M. K. McAllister, B. D. Roitberg, and K. L. Weldon, "Adaptive Suicide in Pea Aphids: Decisions Are Cost Sensitive," *Animal Behaviour*, 40 (1990): 167–175.
18. I. Douglas-Hamilton and O. Douglas-Hamilton, *Among the Elephants* (London: Collins and Harvill, 1975).
19. C. Moss, *Elephant Memories* (Chicago: University of Chicago Press, 2000 [1988]), p. 73.
20. Williams, *Sacred Elephant*; J. Poole, *Coming of Age with Elephants* (London: Hodder & Stoughton, 1996), pp. 156–157.
21. Moss, *Elephant Memories*, pp. 73–74.
22. Moss, *Portraits in the Wild*, p. 33.
23. K. Payne, *Silent Thunder: The Hidden Voice of Elephants* (London: Weidenfeld & Nicolson, 1998), p. 64.
24. T. J. Roper, "Do Badgers Bury Their Dead?" *Journal of Zoology, London*, 234 (1994): 677–680.
25. G. B. Schaller, *The Year of the Gorilla* (Chicago: University of Chicago Press, 1997 [1965]), p. 190; D. Cousins, *The Magnificent Gorilla: The Life History of a Great Ape* (Sussex, U.K.: Book Guild, 1990), p. 263.
26. T. Watkins, personal communication.
27. M. A. Huffman, personal communication. Also described in A. Hofer, M. A. Huffman, and G. Ziesler, *Mahale: A Photographic Encounter with Chimpanzees* (New York: Sterling Publishing, 2000).
28. M. Parker-Pearson, "The Archaeology of Death and Burial" (Texas A & M University Anthropology Series no. 3, 2001 [1999]).

*14. What We Know So Far*

1. D. C. Jarvis, *Folk Medicine: A Doctor's Guide to Good Health* (London: Carnell, 1957), p. 8.

2. B. L. Hart, "Behavioral Adaptations to Pathogens and Parasites: Five Strategies," *Neuroscience and Biobehavioral Reviews,* 14 (1990): 273–294.
3. B. G. Galef, Jr., "Direct and Indirect Behavioural Pathways to the Social Transmission of Food Avoidance," *Annals of the New York Academy of Sciences,* 443 (1984): 203–215.
4. Such examples may warrant the term *zoopharmacognosy* (animal understanding of drugs), while others, in which "understanding" is too grand a word, may be more appropriately described as *zoopharmapractica* (animal interaction with drugs).
5. M. Heinrich, H. Rimpler, and N. A. Barrera, "Indigenous Phytotherapy of Gastrointestinal Disorders in a Lowland Mixed Community (Oaxaca, Mexico) — Ethnopharmacological Evaluation," *Journal of Ethnopharmacology,* 36(1) (1992): 63–80.

### 15. Animals in Our Care

1. E. M. Fowler, *Zoo and Wild Animal Medicine* (Philadelphia: W. B. Saunders, 1993).
2. E. S. Dierenfeld, "Captive Wild Animal Nutrition: A Historical Perspective," plenary lecture, symposium on Nutrition of Wild and Captive Wild Animals, *Proceedings of the Nutrition Society,* 56 (1997): 989–999.
3. D. J. Heard, C. D. Buergelt, et al., "Dilated Cardiomyopathy Associated with Hypovitaminosis E in a Captive Collection of Flying Foxes (*Pteropu* sp.)," *Journal of Zoo and Wildlife Medicine,* 27(2) (1996): 149–157.
4. W. J. Freeland, P. H. Calcott, and L. R. Anderson, "Tannins and Saponins: Interaction in Herbivore Diets," *Biochemial Systematics and Ecology,* 13(2) (1981): 189–193.
5. J. Vermeer, "A Garden of Enrichment," *Animal Keepers Forum,* 5 (1995): 165–167; W. Jens, "Medicinal Herb Use at Apenheul Primate Park," paper presented at EAZA conference, Sept. 1999; R. C. Wren, *Potter's New Cyclopaedia of Botanical Drugs and Preparations* (Saffron Walden: C. W. Daniel, 1988); K. E. Glander, personal communication.
6. J. H. Williams, *Elephant Bill* (London: Rupert Hart-Davis, 1950).
7. M. J. Lynch, R. F. Slocombe, et al., "Fibrous Osteodystrophy in Dromedary Camels (*Camelus dromedus*)," *Journal of Zoology and Wildlife Medicine,* 30(4) (1999): 577–583; J. Hare, director of the Wild Camel Foundation, personal communication.
8. C. K. Morris and V. F. Froelicher, "Cardiovascular Benefits of Physical Activity," *Herz,* 16(4) (1991): 222–236.
9. "Bears Behind Bars," Imago/Animal Planet TV documentary; "Bear Parks in Japan," publication by the World Society for the Protection of Animals, 1999.
10. C. Sugal, *The Price of Habitat* (Washington, D.C.: Worldwatch Institute, 1999); W. C. Mahaney, "Behaviour of the African Buffalo on Mount Kenya," *African Journal of Ecology,* 25 (1987): 199–202.
11. For readers who wish to learn more about natural diets for pets, there is a

wealth of books on the subject, notably those by Juliette de Baïracli Levy, Wendy Volhard, and Richard Peitcairn.

12. J. Hinde, "Saving Polly," *New Scientist* (Sept. 1999): 11.

13. D. Mackensie, "Farm Fresh Food," *New Scientist* (Mar. 18, 2000): 45.

14. Veterinarian S. Kestin, Bristol University, personal communication; R. Guy, director of The Real Meat Company, personal communication; Farm Animal Welfare Council, "Report on the Welfare of Broiler Chickens," *Ministry of Agriculture Fisheries and Food* (Feb. 1999 [Apr. 1992]).

15. "Animal Pharm," *World Animal Health and Nutrition News* (Dec. 8, 1995).

16. Advisory Committee on the Microbiological Safety of Food, *Report on Poultry Meat* (London: HMSO, 1996), p. 49.

17. See review in "Factory Farming and Human Health," *The Ecologist* (Special Report, "How Bogus Hygiene Regulations Are Killing Real Food," June 2001): 30–34.

18. H. R. Kutlu and J. M. Forbes, "Self-selection of Ascorbic Acid in Coloured Foods by Heat-stressed Broiler Chickens," *Physiology and Behaviour,* 53 (1993): 103–110.

19. W. C. Mahaney, B. Maximilliano, et al., "Geophagy of Holstein Hybrid Cattle in the Northern Andes, Venezuela," *Mountain Research and Development,* 16(2) (1996): 177–180; K. L. Clark, A. B. Sarr, et al., "In Vitro Studies on the Use of Clay, Clay Minerals and Charcoal to Adsorb Bovine Rotavirus and Bovine Coronavirus," *Veterinary Microbiology,* 63(2–4) (1998): 137–146; U.S. Department of Agriculture's National Animal Health Monitoring Service (NAHMS) 1992 on-line database.

20. J. Cooper, I. J. Gordon, and A. W. Pike, "Strategies for the Avoidance of Faeces by Grazing Sheep," *Applied Animal Behaviour Science,* 69 (2000): 15–33; G. H. Scales, T. L. Knight, and D. J. Saville, "Effect of Herbage Species and Feeding Level on Internal Parasites and Production Performance of Grazing Lambs," *New Zealand Journal of Agricultural Research,* 38 (1994): 237–247; R. J. Aerts, T. N. Barry, and W. C. McNabb, "Polyphenols and Agriculture: Beneficial Effects of Proanthocyanidins in Forages," *Agriculture Ecosystems and Environment,* 75(1–2) (1999): 1–12; J. H. Niezen, W. A. G. Charleston, et al., "Controlling Internal Parasites in Grazing Ruminants Without Recourse to Anthelmintics: Approaches, Experiences and Prospects," *International Journal of Parasitology,* 26(8–9) (1996): 983–992.

21. A. F. Fraser, "Animal Suffering: The Appraisal and Control of Depression and Distress in Livestock," *Applied Animal Behaviour Science,* 20 (1988): 127–133.

## 16. Healthy Intentions

1. National Center for Health Statistics, 1998; data supplied by The Samaritans (U.K.) and the World Health Organization.

2. S. B. Eaton, M. Shostak, and M. Konner, *The Paleolithic Prescription: A Program of Diet and Exercise and a Design for Living* (New York: Harper & Row, 1988); J. Challem, "Paleolithic Nutrition: Your Future Is in Your Dietary Past," *Nutrition Science News* (Apr. 1997).

3. R. M. Nesse and G. C. Williams, *Why We Get Sick* (New York: Vintage Books, 1996), p. 9.

4. T. Johns, *The Origins of Human Diet and Medicine* (Tucson: University of Arizona Press, 1993).

5. D. G. Popovich, D. J. A. Jenkins, et al., "The Western Lowland Gorilla Diet Has Implications for the Health of Humans and other Hominoids," *Human and Clinical Nutrition,* 127 (1997): 2000–5.

6. M. H. Logan and A. R. Dixon, "Agriculture and the Acquisition of Medicinal Plant Knowledge," in *Eating on the Wild Side,* ed. N. L. Etkin (Tucson: University of Arizona Press, 1994), p. 29.

7. G. J. Armelagos, A. H. Goodman, and K. H. Jacobs, "The Origin of Agriculture: Population Growth During a Period of Declining Health," *Population and Environment,* 13(1) (1991): 9–22; H. C. Trowell and D. P. Burkitt, *Western Diseases: Their Emergence and Prevention* (Cambridge, Mass.: Harvard University Press, 1981).

8. United Nations Environment Program, *Global BioDiversity: Earth's Living Resources in the Twenty-first Century* (UN Publications, 2000).

9. T. Johns, "Plant Constituents and the Nutrition and Health of Indigenous Peoples," in *Ethnoecology,* ed. V. D. Nazarea (Tucson: University of Arizona Press, 1999), pp. 157–174; T. Johns, "The Chemical Ecology of Human Ingestive Behaviour," *Annual Review of Anthropology,* 28 (1999): 27–47.

10. D. Hensrud and D. Heimburger, "Diet, Nutrients and Gastrointestinal Cancer," *Gastroenterology Clinics of North America,* 27(2) (1998): 325–348.

11. W. Craig and L. Beck, "Phytochemicals: Health and Protective Effects," *Canadian Journal of Dietetic Practise and Research,* 60(2) (1999): 78–84.

12. R. M. Sapolsky, *Junk Food Monkeys* (London: Headline, 1998), pp. 103–112.

13. Johns, "Plant Constituents."

14. R. N. Hughes, *Diet Selection: An Interdisciplinary Approach to Optimal Foraging* (London: Blackwell Scientific Press, 1993), pp. 2–5; Nesse and Williams, *Why We Get Sick,* p. 30; G. Maciocia, *The Foundations of Chinese Medicine* (London: Churchill Livingstone, 1989), p. 276; E. J. Khantzian, "The Self-medication Hypothesis of Substance Use Disorders: A Reconsideration and Recent Applications," *Harvard Review of Psychiatry,* 4(5) (1997): 231–244.

# INDEX